Eye and Brain

Third edition

Frontispiece The visual region of the brain – the *area striata*.
Here we see a small part of the mechanism of the brain
highly magnified.,These cell bodies with their connections
handle information from the eyes, to give us knowledge
of the world.

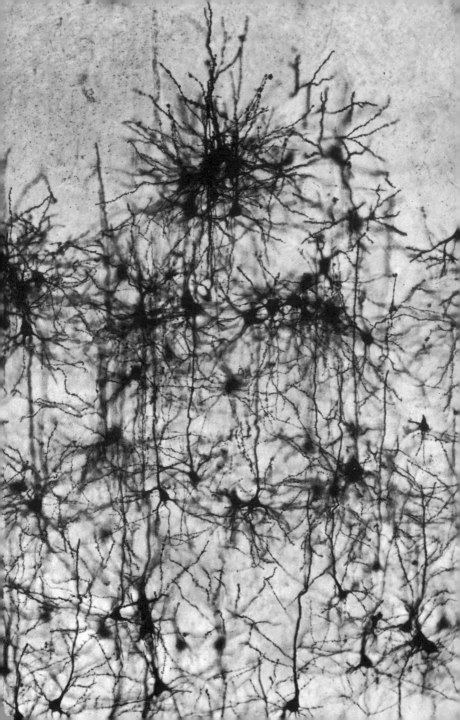

Eye and Brain

The psychology of seeing

R.L.Gregory

Weidenfeld and Nicolson
London

First published March 1966
Second impression June 1966
Third impression January 1967
Fourth impression March 1969
Fifth impression September 1970
Second edition 1972
Third edition 1977

Weidenfeld and Nicolson
11 St John's Hill, London SW11 1XA

Filmset by BAS Printers Limited, Wallop, Hampshire
Manufactured by Librex, Milan, Italy

ISBN 297 77298 8 cased
ISBN 297 77303 8 paperback

Contents

Foreword to the Third Edition

May I take this opportunity to thank the publishers for producing in the first place, and now revising, *Eye and Brain*. Appearing in 1966, it was the first volume of the World University Library. This was the imaginative conception of Lord Weidenfeld. Production could be made quite lavish, with many colour pictures, by printing in several countries and as many languages simultaneously. The more financially successful books of the series supported the more specialised, which could not be expected to command economic sales with the production standards of this conception. The series was originally edited by Colin Haycraft, who has remained a close and valued friend.

Research on eyes and brains – on how our predecessors and ourselves on the incredible ladder of life see and understand – has made exciting advances during the ten years since this book was first written. I can only hope that some of this has been captured and encapsulated in this edition. I have also taken this opportunity to rephrase passages which seemed less than happy.

May I thank particularly John Curtis. He was responsible for editing – and made immense contributions to – *The Intelligent Eye*, which appeared in 1970. With his usual care and imaginative judgment, he has now produced this new edition of *Eye and Brain*.

R.L.G.

1 Seeing

We are so familiar with seeing, that it takes a leap of imagination to realise that there are problems to be solved. But consider it. We are given tiny distorted upside-down images in the eyes, and we see separate solid objects in surrounding space. From the patterns of stimulation on the retinas we perceive the world of objects, and this is nothing short of a miracle.

The eye is often described as like a camera, but it is the quite uncamera-like features of perception which are most interesting. How is information from the eyes coded into neural terms, into the language of the brain, and reconstituted into experience of surrounding objects? The task of eye and brain is quite different from either a photographic or a television camera converting objects merely into images. There is a temptation, which must be avoided, to say that the eyes produce pictures in the brain. A picture in the brain suggests the need of some kind of internal eye to see it – but this would need a further eye to see *its* picture . . . and so on in an endless regress of eyes and pictures. This is absurd. What the eyes do is to feed the brain with information coded into neural activity – chains of electrical impulses – which by their code and the patterns of brain activity, represent objects. We may take an analogy from written language: the letters and words on this page have certain meanings, to those who know the language. They affect the reader's brain appropriately, but they are not pictures. When we look at something, the pattern of neural activity represents the object and to the brain *is* the object. No internal picture is involved.

Gestalt writers did tend to say that there are pictures inside the brain. They thought of perception in terms of modifications of electrical fields of the brain, these fields copying the form of perceived objects. This doctrine, known as isomorphism, has had unfortunate effects on thinking about perception. Ever since, there has been a tendency to postulate properties to these hypothetical brain fields such that visual distortions, and other phenomena, are 'explained'.

But it is all too easy to postulate things having just the right properties. There is no independent evidence for such brain fields, and no independent way of discovering their properties. If there is no evidence for them, and no way of discovering their properties, then they are highly suspect. Useful explanations relate observables.

The Gestalt psychologists did however point to several important phenomena. They also saw very clearly that there is a problem in how the mosaic of retinal stimulation gives rise to perception of objects. They particularly stressed the tendency for the perceptual system to group things into simple units. This is seen in an array of dots (figure 1.1). Here the dots are in fact equally spaced, but there is a tendency to see, to 'organise', the columns and rows as though they are separate objects. This is worth pondering, for in this example we have the essential problem of perception. We can see in ourselves the groping towards organising the sensory data into objects. If the brain were not continually on the look-out for objects, the cartoonist would have a hard time. But, in fact, all he has to do is present a few lines to the eye and we see a face, complete with an expression. The few lines are all that is required for the eye – the brain does the rest: seeking objects and finding them whenever possible. Sometimes we see objects which are not there: faces-in-the-fire, or the Man in the Moon.

Figure 1.2 is a joke figure which brings out the point clearly. Just a set of meaningless lines? No – it is a washer-woman with her bucket! Now look again: the lines are subtly different, almost solid – they represent objects.

The seeing of objects involves many sources of information beyond those meeting the eye when we look at an object. It generally involves knowledge of the object derived from previous experience, and this experience is not limited to vision but may include the other senses; touch, taste, smell, hearing and perhaps also temperature or pain. Objects are far more than patterns of stimulation: objects have pasts and futures; when we know its past or can guess its future, an object transcends experience and becomes an embodiment of knowledge and expectation without which life beyond the simplest is not possible.

Although we are concerned with how we see the world of objects, it is important to consider the sensory processes giving perception –

1.1 This array of equally spaced dots is seen as continually changing patterns of rows and squares. We see something of the active organising power of the visual system while looking at this figure.

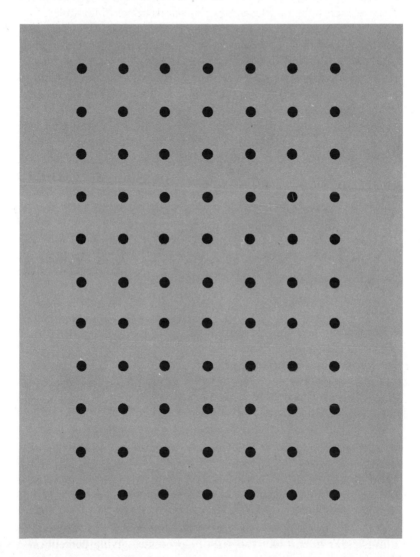

1.2 A joke figure – what is it? When you see it as an object, not merely meaningless lines, it will suddenly appear almost solid – an *object*, not a *pattern*.

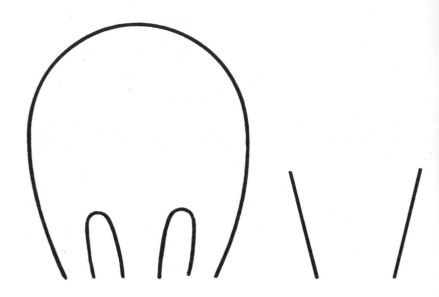

what they are, how they work and when they fail to work properly. It is by coming to understand these underlying processes that we can understand how we perceive objects.

There are many familiar so-called 'ambiguous figures', which illustrate very clearly how the same pattern of stimulation at the eye can give rise to different perceptions, and how the perception of objects goes beyond sensation. The most common ambiguous figures are of two kinds: figures which alternate as 'object' or 'ground', and those which spontaneously change their position in depth. Figure 1.3 shows a figure which alternates in figure and ground – sometimes the black part appears as a face, the white being neutral background, and at others the black is insignificant and the white surround dominates and seems to represent an object. The well-known Necker cube (figure 1.4) shows a figure alternating in depth. Sometimes the face

12

1.3 This figure alternates spontaneously, so that sometimes it is seen as a pair of faces, sometimes as a white urn bounded by meaningless black areas – the faces. The perceptual 'decision' of what is figure (or object) and what ground, is similar to the engineer's distinction between 'signal' and 'noise'. It is basic to any system which handles information.

marked with the 'o' lies in front, sometimes at the back – it jumps suddenly from the one position to the other. Perception is not determined simply by the stimulus patterns; rather it is a dynamic searching for the best interpretation of the available data. The data are sensory signals, and also knowledge of the many other characteristics of objects. Just how far experience affects perception, how far we have to learn to see, is a difficult question to answer; it is one which will concern us in this book. But it seems clear that perception involves going beyond the immediately given evidence of the senses: this evidence is assessed on many grounds and generally we make the best bet, and see things more or less correctly. But the senses do not give us a picture of the world directly; rather they provide evidence for the checking of hypotheses about what lies before us. Indeed, we may say that the perception of an object *is* an hypothesis, suggested

13

1.4 This figure alternates in depth: the face of the cube marked by the small circle sometimes appearing as the *front*, sometimes as the *back* face.
We can think of these ways of seeing the figure as perceptual 'hypotheses'. The visual system entertains alternative hypotheses, and never settles for one solution. This process goes on throughout normal perception, but generally there is a unique solution.

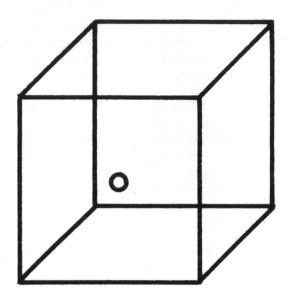

and tested by the sensory data. The Necker cube is a pattern which contains no clue as to which of two alternative hypotheses is correct: the perceptual system entertains first one then the other hypothesis, and never comes to a conclusion, for there is no best answer. Sometimes the eye and brain come to wrong conclusions, and then we suffer hallucinations or illusions. When a perceptual hypothesis – a perception – is wrong we are misled, as we are misled in science when we see the world distorted by a false theory. Perceiving and thinking are not independent: 'I see what you mean' is not a puerile pun, but indicates a connection which is very real.

2 Light

To see, we need light. This may seem too obvious to mention but it has not always been so obvious – Plato thought of vision as being due not to light entering, but rather to particles shot out of the eyes, spraying surrounding objects. It is difficult to imagine now why Plato did not try to settle the matter with a few simple experiments. Although to philosophers the problem of how we see has always been a favourite topic of speculation and theory, it is only in the last hundred years that it has formed the object of systematic experiments; which is odd, because all scientific observations depend upon the human senses – most particularly upon sight.

For the last three hundred years there have been two rival theories of the nature of light. Isaac Newton (1642–1727) argued that light must be a train of particles, while Christiaan Huygens (1629–93) argued that it must be pulses travelling through an all-pervading medium – the *aether* – which he thought of as small elastic balls in contact with each other. Any disturbance, he suggested, would be carried in all directions through the packed spheres as a wave, and this wave is light.

The controversy over the nature of light is one of the most exciting and interesting in the history of science. A crucial question in the early stages of the discussion was whether light travelled at a finite speed, or whether it arrived instantaneously. This was answered in a quite unexpected way by a Danish astronomer Olaus Roemer (1644–1710). He was engaged in recording eclipses of the four bright satellites orbiting round Jupiter, and found that the times he observed were not regular, but depended upon the distance of Jupiter from the earth.

He came to the conclusion, in 1675, that this was due to the time light took to reach him from the satellites of Jupiter, the time increasing when the distance increased, because of the finite velocity of light. In fact, the distance of Jupiter varies by about 300,000,000 km. (twice the distance of the sun), and the greatest time-

difference he observed was 16 minutes 36 seconds earlier or later than the calculated time of the eclipses of the satellites. From his somewhat faulty estimate of the distance of the sun he calculated the speed of light as 309,000 km. per second. With our modern knowledge of the diameter of the earth's orbit, we correct this to a velocity of about 300,000 km. per second, or 3×10^{10} cm/sec. The speed of light has since been measured very accurately over short distances on earth, and it is now regarded as one of the basic constants of the Universe.

Because of the finite velocity of light, and the delay in nervous messages reaching the brain, we always sense the past. Our perception of the sun is over eight minutes late: all we know of the furthest object visible to the unaided eye (the Andromeda nebula) is so out of date that we see it as it was a million years before men appeared on Earth.

The value of 3×10^{10} cm/sec for the speed of light strictly holds only for a perfect vacuum. When light travels through glass or water, or any other transparent substance, it is slowed down to a velocity which depends upon the refractive index (roughly the density) of the medium through which it is travelling. This slowing down of light is extremely important, for it is this which causes prisms to bend light, and lenses to form images. The principle of refraction (the bending of light by changes of refractive index) was first understood by Snell, a Professor of Mathematics at Leyden, in 1621. Snell died at thirty-five, leaving his results unpublished. René Descartes (1596–1650) published the Law of Refraction eleven years later. The Law of Refraction (The 'Sine Law') is:

When light passes from a medium A into a medium B the sine of the angle of incidence bears to the sine of the angle of refraction a constant ratio.

We can see what happens with a simple diagram (figure 2.3): if AB is a ray passing from a dense medium (glass) into a vacuum (or air) the ray will emerge into the air at some angle i along BD.

2.2 Sir Isaac Newton (1642–1727) by Charles Jervas. On the whole, Newton held that light consists of particles, but he was aware of many of the difficulties, anticipating the modern theory that light has the dual properties of particles and waves. He devised the first experiments to show that white light is a mixture of the spectral colours, and paved the way to an understanding of colour vision by elucidating the physical characteristics of light.

The Law states that $\frac{\sin i}{\sin r} = \mu$. The constant μ is the refractive index of the glass or other refracting substance.

Newton thought of his corpuscles of light as being attracted to the surface of the denser medium, while Huygens thought that the bending was due to the light travelling more slowly in the denser medium. It was many years before the French physicist Foucault showed by direct measurement that light does indeed travel more slowly in a denser medium. It seemed for a time that Newton's corpuscle theory of light was entirely wrong – that light is purely a series of waves radiating through a medium, the *aether* – but at the beginning of the present century it was dramatically shown that the wave theory does not explain all the phenomena of light. It now seems that light is both corpuscles and waves.

Light consists of packets of energy – *quanta* – these combining the characteristics of corpuscles and waves. Light of short wavelength has more waves in each bundle than light of longer wavelength. This is expressed by saying that the energy of a single quantum is a function of frequency, such that $E = h\nu$ where E is the energy in ergs, h is a small constant (Planck's constant) and ν is the frequency of the radiation.

When light is bent by a prism, each frequency is deviated through a slightly different angle, so that the emergent beam comes out of the prism as a fan of light, giving all the spectral colours. Newton discovered that white light is a compound of all the spectral colours, by splitting a beam of sunlight into a spectrum in this way and then finding that he could re-combine the colours back into white light by passing the spectrum through a second similar prism, held the other way up. Newton named seven colours for his spectrum – red, orange, yellow, green, blue, indigo, violet. One does not really see indigo as a separate colour, and orange is a bit doubtful. What happened is that Newton liked the number seven and added the names orange and indigo to make the magic number!

18

D. ISAACVS NEWTON EQVES
REG. SOCIETATIS PRÆSES. AN.̊ 1702.

2.3 Light is bent (refracted) by a dense transparent medium. The ratio of the sines of the angles of the rays entering and leaving the dense medium are constant for a given refractive index of the medium. This is the basis of image formation by lenses. (The angle of deviation of light is also a function of the wavelength of light, so that a beam is split into the spectral colours by a prism.) The lettering is explained in the text.

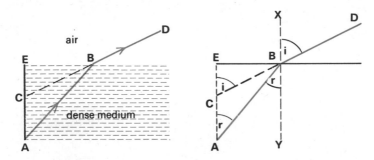

We know now, though Newton did not, that each spectral colour, or hue, is light of a different frequency. We also know that all so-called electromagnetic radiation is essentially the same. The physical difference between radio waves, infra-red, light, ultraviolet and X-rays is their frequency. Only a very narrow band of these frequencies, less than an octave in width, stimulates the eye to give vision and colour. The diagram (figure 2.5) shows how this narrow window fits into the physical picture. Looked at in this way, we are almost blind.

If we know the speed of light and its frequency, it is a simple matter to calculate its wavelength, but in fact its frequency is difficult to measure directly. It is easier to measure the wavelength of light than its frequency; though this is not so for the low frequency radio waves. The wavelength of light is measured by splitting it up not with a prism, but with a grating of finely ruled lines, which also produces the colours of the spectrum. (This can be seen by holding an L.P. record at an oblique angle to a source of light, when the reflection will be made up of brilliant colours.) Given the spacing of the lines of a grating, which are specially ruled, and the angle producing light of a given colour, wavelength may be determined very accurately. It turns out that the blue light has a wavelength of about 4×10^{-5} cm. ($\frac{1}{70,000}$ inch), while the wavelength of red light is about 7×10^{-5} cm. ($\frac{1}{40,000}$ inch). The range of wavelengths adopted by eyes is important for it

2.4 A freehand sketch by Newton of one of his experiments on colour.
He first split light into a spectrum (with the large prism), then allowed
light of a single colour to pass through a hole in a screen to a second prism.
This did not produce more colours. He also found that a second prism
placed in the spectrum would recombine the colours into white.
Thus white light is made up of all the colours of the spectrum.

sets the limit to their resolution, just as for optical instruments such as
cameras, and it is adapted to accept the wavelengths of maximum
energy of sunlight as it is filtered by the atmosphere.

We cannot with the unaided eye see individual quanta of light, but
the receptors in the retina are so sensitive that they can be stimulated
by a single quantum, though several (five to eight) are required to give
the experience of a flash of light. The individual receptors of the retina
are as sensitive as it is possible for any light detector to be, since a
quantum is the smallest amount of radiant energy which can exist. It
is rather sad that the transparent media of the eye do not quite match
this development to perfection. Only about ten per cent of the light
reaching the eye gets to the receptors, the rest being lost by absorption
and scattering within the eye before the retina is reached. In spite of

21

this loss, it would be possible under ideal conditions to see a single candle placed seventeen miles away.

The quantal nature of light has an important implication for vision, which has inspired some particularly elegant experiments bridging the physics of light and its detection by the eye and brain. The first experiment on the effect of light being packaged into quanta was undertaken by three physiologists, Hecht, Schlaer, and Pirenne, in 1942. Their paper is now a classic. Realising that the eye must be almost if not quite as sensitive as theoretically possible, they devised a most ingenious experiment for discovering how many quanta actually accepted by the receptors are required to see a flash of light. The argument is based on a statistical function known as the *Poisson distribution*. This gives the expected distribution of hits on a target. The idea is that part at least of the moment-to-moment variation in the effective sensitivity of the eye is not due to anything in the eye or the nervous system, but to the variation in moment-to-moment energy of weak light sources. Imagine a desultory rain of bullets: they will not arrive at a constant rate, but will fluctuate; similarly there is fluctuation in the number of light quanta that arrive. A given flash may contain a small or large number of quanta, and is more likely to be detected if there happen to be more than the average number of quanta in the flash. For bright lights, this effect is unimportant, but since the eye is sensitive to but a few quanta, the fluctuation is sufficient for estimating the number of quanta required for detection.

The quantal nature of light is also important in considering the ability of the eye to detect fine detail. One of the reasons why it is possible to read only the largest newspaper headlines by moonlight, is that there are insufficient quanta falling onto the retina to build up a complete image within the time-span over which they eye can integrate energy – about a tenth of a second. In fact this is not by any means the whole story; but the purely physical factor of the quantal nature of light contributes to a well known visual phenomenon – loss of acuity in dim light – which until recently has been treated purely as though it were a property of the eye. Indeed, it is often quite difficult to establish whether a visual effect should be regarded as belonging to psychology, physiology or physics. They get pretty well mixed up.

How are images produced? The simplest way an image can be formed is by a pin hole. Figure 2.6 shows how this comes about. A ray

2·5 Light is but a narrow region of the total electromagnetic spectrum, which includes radio waves, infra red, ultra violet and X-rays.
The physical difference is purely the wavelength of the radiation, but the effects are very different. Within the octave to which the eye is sensitive, different wavelengths give different colours. Beyond light these are very different properties when radiation interacts with matter.

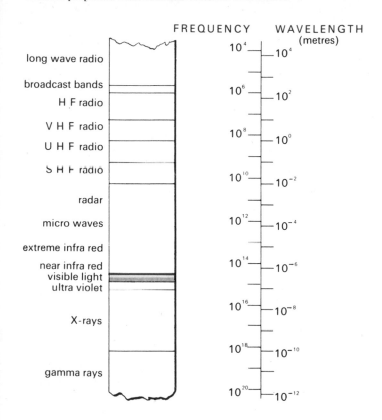

How are images produced? The simplest way an image can be formed is by a pin hole. Figure 2·6 shows how this comes about. A ray from a part of the object (x) can only reach one part of the screen (y) – the part lying along the straight line passing through the pin-hole. Each part of the object illuminates a corresponding part of the screen, so an upside-down picture of the object is formed on the screen. The pin-hole image will be rather dim, for

2.6 Forming an image with a pinhole. A ray from a given region of the source reaches only a single region of the screen – the ray passing through the hole. Thus an (inverted) image is built up from the rays lying in the path of the hole. The image is free from distortion, but is dim and not very sharp. A very small hole introduces blurring through diffraction effects, due to the wave nature of light.

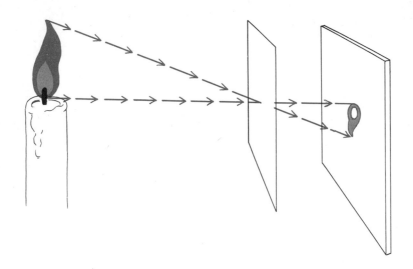

from a part of the object (x) can only reach one part of the screen (y) – the part lying along the straight line passing through the pin hole. Each part of the object illuminates a corresponding part of the screen, so an upside-down picture of the object is formed on the screen. The pin hole image will be rather dim, for the hole must be small if the image is to be sharp. (Though if it is *too* small it will be blurred because the wave structure of the light is upset).

A lens is really a pair of prisms (figure 2.7). It directs a lot of light from each point of the object to a corresponding point on the screen, thus giving a bright image. Lenses only work well when they are suitable, and adjusted correctly. The lens of the eye can be unsuitable to the eye in which it finds itself, and it can be adjusted wrongly. The lens may focus the image in front of or behind the retina, instead of on it, giving 'short' or 'long' sight. The lens may not be truly spherical in its surface, giving distortion and, in some directions, blurring of the

2.7 A lens can be thought of as a pair of converging prisms, forming an image from a bundle of rays. The image is far brighter than from a pinhole, but it is generally distorted in some degree, and the depth of focus is limited. [This figure should not be taken too literally – image-forming lenses have curved surfaces.]

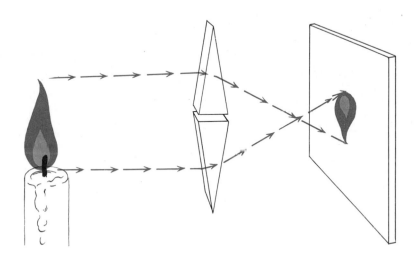

image. The cornea may be irregular, or pitted (perhaps due to abrasion from metal filings in industry, or grit from riding a motor cycle without protective goggles). Most optical defects can be corrected by adding artificial lenses – spectacles. Spectacles correct for errors in accommodation by changing the power of the lens of the eye; they correct for astigmatism by adding a non-spherical component. Ordinary spectacles cannot correct for damage to the surface of the cornea but *corneal lenses*, fitted to the eye itself, serve to give a fresh surface to the cornea.

Spectacles lengthen our effective lives. With their aid we can see to read and to perform skilled tasks into old age: before their invention scholars and craftsmen were made helpless by lack of sight, though they still had the power of their minds.

3.1 Various primitive eyes. All of the ones here have the same basic plan: a lens forming an image on a mosaic of light-sensitive receptors.

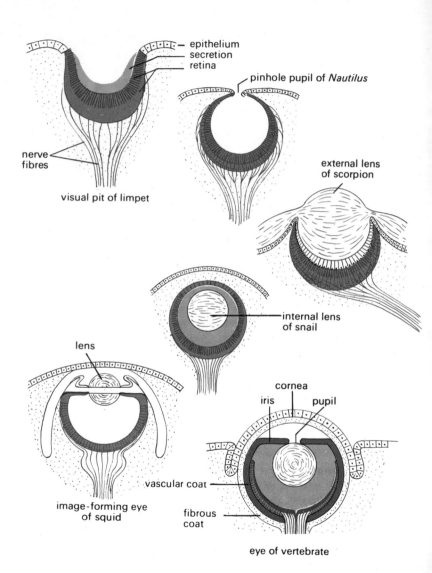

epithelium
secretion
retina

pinhole pupil of *Nautilus*

external lens of scorpion

nerve fibres

visual pit of limpet

internal lens of snail

lens

cornea
iris   pupil

image-forming eye of squid

vascular coat

fibrous coat

eye of vertebrate

3 In the beginning . . .

Almost every living thing is sensitive to light. Plants accept the energy of light, some moving to follow the sun almost as though flowers were eyes to see it. Animals make use of light, shadows, and images to avoid danger and to seek their prey.

The first simple eyes responded only to light, and changing intensity of light. Perception of form and colour waited upon more complicated eyes capable of forming images, and brains sufficiently elaborate to interpret the neural signals from optical images on the retinas.

The later image-forming eyes developed from light-sensitive spots on the surface of simpler animals. How this occurred is largely mysterious, but we know some of the characters in the story. Some can be seen as fossils; some are inferred from comparative studies of living species; others appear fleetingly during the development of embryo eyes.

The problem of how eyes have developed has presented a major challenge to the Darwinian theory of evolution by Natural Selection. We can make many entirely useless experimental models when designing a new instrument, but this was impossible for Natural Selection, for each step must confer some advantage upon its owner, to be selected and transmitted through the generations. But what use is a half-made lens? What use is a lens giving an image, if there is no nervous system to interpret the information? How could a visual nervous system come about before there was an eye to give it information? In evolution there can be no master plan, no looking ahead to form structures which, though useless now, will come to have importance when other structures are sufficiently developed. And yet the human eye and brain have come about through slow painful trial and error.

Response to light is found even in one-celled animals. In higher forms, we find specially adapted cells to serve as receptors sensitive to light. These cells may be scattered over the skin (as in the earth worm)

3.2 The fossil eye of a species of trilobite. This is the earliest kind of eye preserved as a fossil. The facets are the corneal lenses, essentially the same as a modern insect eye. Some trilobites could see all round, but none above.

or they may be arranged in groups, most often lining a depression or pit, which is the beginning of a true image-forming eye.

It seems likely that photoreceptors became recessed in pits because there they lay protected from the surrounding glare, which reduced their ability to detect moving shadows heralding the approach of danger. Millions of years later but for the same reason Greek astronomers, it is said (though this has been questioned), dug pits in the ground from which they observed stars in the daytime.

The primitive eye pits were open to the danger of becoming blocked by foreign particles lodging within them, shutting out the light. A transparent protective membrane developed over the eye pits, serving to protect them. When, by chance mutations, this membrane became thicker in its centre, it became a crude lens. The first lenses served merely to increase intensity, but later they came to form useful images. An ancient pit type of eye is still to be seen in the limpet. One living creature, *Nautilus*, has an eye still more primitive – there is no lens, but a pin hole to form the image. The inside of the eye of *Nautilus* is washed by the sea in which it lives, while eyes with lenses are filled with specially manufactured fluids to replace the sea. Human tears are a re-creation of the primordial ocean, which bathed the first eyes (figure 3.1).

We are concerned in this book with human eyes, and how we see the world. Our eyes are typical vertebrate eyes, and are not among the most complex or highly developed, though the human brain is the most elaborate of all brains. Complicated eyes often go with simple brains – we find eyes of incredible complexity in pre-vertebrates serving tiny brains. The compound eyes of arthropods (including insects) consist not of a single lens with a retina of many thousands or millions of receptors, but rather of many lenses with a set of about seven receptors, for each lens. The earlier known fossil eye belongs to the trilobites, which lived 500,000,000 years ago. In many species of trilobite, the eyes were highly developed. The external structure of these most ancient eyes may be seen perfectly preserved (figure 3.2). Some of the internal structure can be seen in the fossils with X-rays. They were compound eyes, rather like those of a modern insect: some had over a thousand facets.

Figure 3.3 shows an insect eye. Behind each lens facet ('corneal lens') lies a second lens ('lens cylinder') through which light passes to the light-sensitive element, this usually consisting of seven cells grouped in a minute flowerlike cluster. Each complete unit of a compound eye is known as an 'ommatidium'. It used to be thought that each ommatidium is a separate eye – so that insects must see thousands of worlds – but how this came to be believed is strange, for there is no separate retina in each ommatidium, and but a single nerve fibre from each little group of receptors. How then could each one signal a complete image? The fact is that each ommatidium signals the presence of light from a direction immediately in front of it, and the combined signals represent effectively a single image.

Insect eyes have a remarkable mechanism to give dark or light adaptation. The ommatidia are isolated from each other by black cones of pigment: with reduced light (or in response to signals from the brain) the pigment migrates back towards the receptors so that light can then pass through the side of each ommatidium to neighbouring receptors. This increases the sensitivity of the eye, but at a cost to its acuity – a trade balance found also in vertebrate eyes, though for somewhat different reasons, and by very different mechanisms.

The lens cylinder of the compound eye does not function by virtue of the shape of its optical surfaces, as in a normal lens, so much as by change of its refractive index, which is greater near its centre than at its edge. Light is funnelled through it, in a way quite different from a normal lens. Compound eyes are principally detectors of movement, and can be incredibly efficient, as we known from watching a dragon fly catching its prey on the wing.

Among the most curious eyes in the whole of nature is that of a creature the size of a pin's head – a little known copepod – *Copilia*. She (the males are dull by comparison) has a pair of image-forming eyes, which function neither like vertebrate nor like compound eyes, but something like a television camera. Each eye contains two lenses, and the photoreceptor system is similar to the insect eye; but in *Copilia* there is an enormous distance between the corneal lens and the lens cylinder. Most of the eye lies deep within the body of the animal, which is extraordinarily transparent. She is shown in figure 3.4. The secret of this eye is to be found by looking at the living animal. Exner, in 1891, reported that the receptor (and attached lens cylinder) make a 'continuous lively motion.' They oscillate across the

3.3 The parts of a compound eye. The primitive trilobite eye was probably similar, though the internal structure is not preserved. We see this in arthropods, including insects, e.g. bees and dragon flies. Each corneal lens provides a separate image to a single functional receptor (made up often of seven light-sensitive cells), but there is no reason to think that the creature sees a mosaic. The compound eye is especially good at detecting movement.

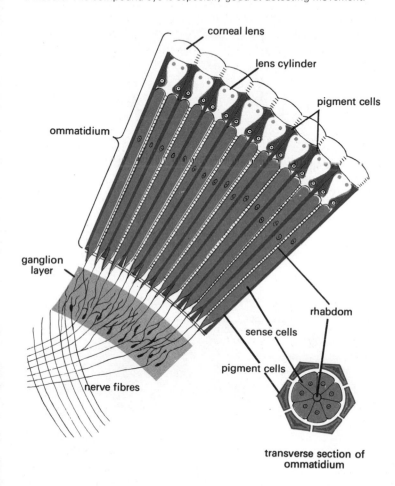

transverse section of
ommatidium

3.4 A living female specimen of a microscopic copepod, *Copilia quadrata*. Each eye has two lenses: a large anterior lens and a second smaller lens deep in the body, with an attached photoreceptor and single optic nerve fibre to the central brain. The second lens and photoreceptor are in continual movement, across the image plane of the first lens. This seems to be a scanning eye: a mechanical television camera.

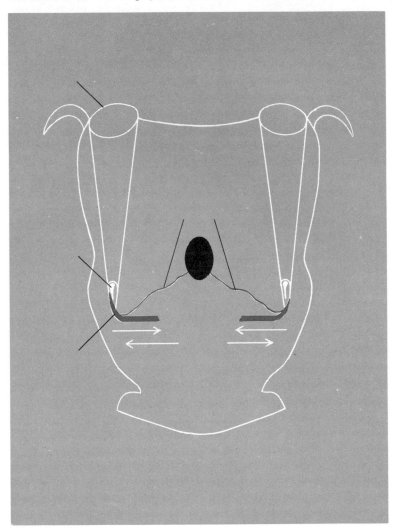

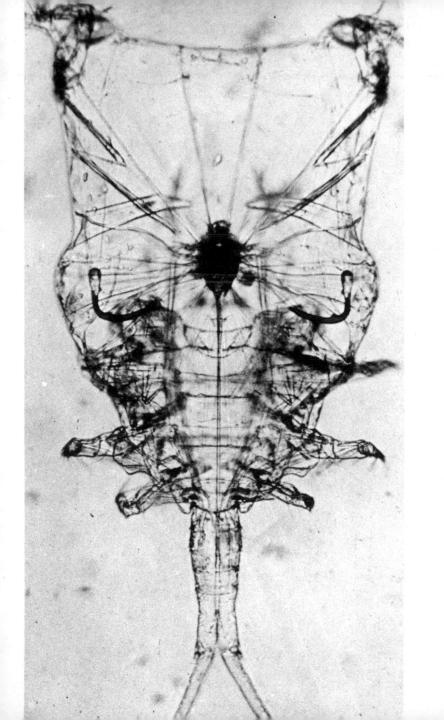

3.5 The posterior lens of *Copilia*, and attached photoreceptor (in red) during a single scan. The rate can be as high as 5 scans per second.

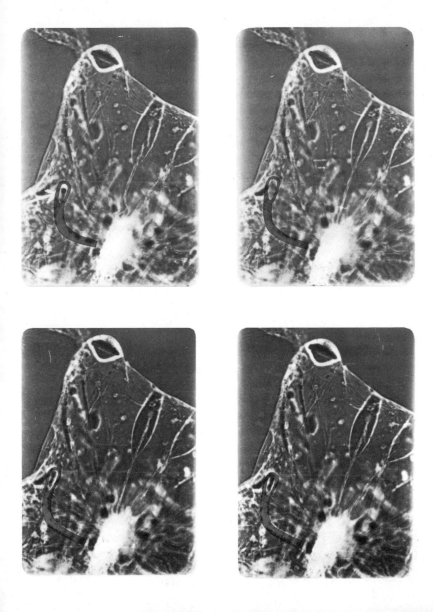

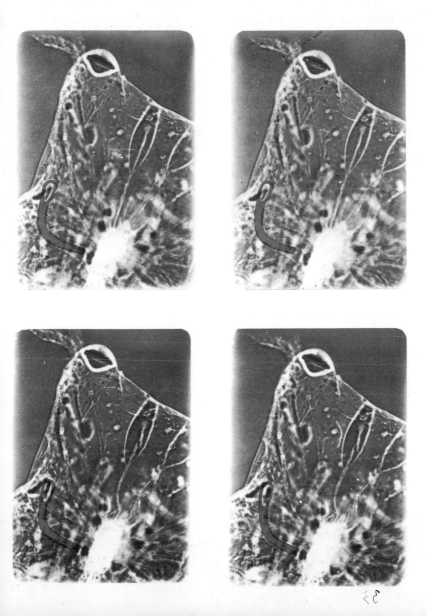

mid line of the animal, and evidently scan across the focal plane of the front corneal lens. It seems that the pattern of dark and light of the image is not given simultaneously by many receptors, as in other eyes, but in a time-series down the optic nerve, as in the single channel of a television camera. It is possible that many small compound eyes (e.g. *Daphnia*) also go in for scanning to improve the resolution and channel capacity of their few elements. Is *Copilia*'s eye ancestral to the compound eye? Is scanning generally abandoned, because a single neural link could not transmit sufficient information? Is it a simplification of the compound eye found in the earliest fossils? Or is it perhaps an aberrant experiment, unrelated to the main streams of evolutionary development?

The scanning movement of the lens cylinder and the attached photoreceptor is shown by the successive frames of a cine film in figure 3.5. The receptors move precisely towards, then away from each other – never independently. The speed of the scan varies from about five per second to about one scan every two seconds. One would give a lot to know why it exists, and whether it is the remaining example of a very early kind of eye. If *Copilia* is an evolutionary failure she deserves a prize for originality.

4 The brain

The brain is more complicated than a star and more mysterious. Looking, with imagination, back through the eyes to the brain mechanisms lying behind, we may there discover secrets as important as the nature of the external world, perceived by eye and brain.

It has not always been obvious that the brain is concerned with thinking, with memory, or with sensation. In the ancient world, including the great civilisations of Egypt and Mesopotamia, the brain was regarded as an unimportant organ. Thought and the emotions were attributed to the stomach, the liver and the gall bladder. Indeed, the echo lingers in modern speech, in such words as 'phlegmatic'. When the Egyptians enbalmed their dead they did not trouble to keep the brain (which was extracted through the left nostril) though the other organs were separately preserved, in special Canopic jars placed beside the sarcophagus. In death the brain is almost bloodless, so perhaps it seemed ill-suited to be the receptacle of the Vital Spirit. The active pulsing heart seemed to be the seat of life, warmth and feeling – not the cold grey silent brain, locked privily in its box of bone.

The vital role of the brain in control of the limbs, in speech and thought, sensation and experience, gradually became clear from the effects of accidents in which the brain was damaged. Later the effects of small tumours and gun shot wounds gave information which has been followed up and studied in great detail. The results of these studies are of the greatest importance to brain surgeons: for while some regions are comparatively safe, others must not be disturbed or the patient will suffer grievous loss.

The brain has been described as 'the only lump of matter we know from the inside.' From the outside, it is a pink-grey object, about the size of two clenched fists. Some relevant parts are shown in figure 4.1. It is made up of the so-called 'white' and 'grey' matter, the white matter being association fibres, connecting the many kinds of cell bodies which form the grey matter.

4.1 The brain, showing the visual part – the *area striata* – at the back (occipital cortex). Stimulation of small regions produces flashes of light in corresponding parts of the visual field. Stimulation of surrounding regions (visual association areas) produces more elaborate visual experiences.

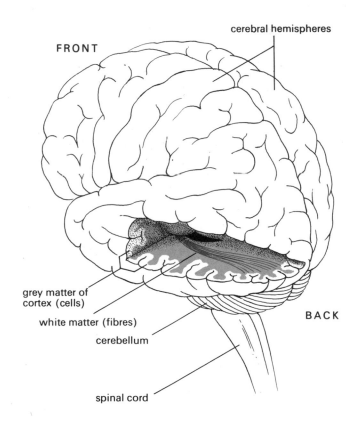

cerebral hemispheres

FRONT

grey matter of
cortex (cells)

white matter (fibres)

cerebellum

BACK

spinal cord

The brain has, in its evolution, grown up from the centre, which in man is concerned primarily with emotion. The surface – the *cortex* – is curiously convoluted. It is largely concerned with motor control of the limbs, and with the sense organs. It is possible to obtain maps relating regions of the cortex with groups of muscles; and also maps relating touch – giving bizarre 'homunculi' as in figure 4.2. The sense of sight has its own region of cortex, as we shall see in a moment.

4.2 A 'homunculus' – a graphic representation showing how much of the cortex is devoted to sensation from various regions of the body. Note the huge thumb. Different animals have very different 'homunculi', corresponding to the sensory importance of the various parts of the body.

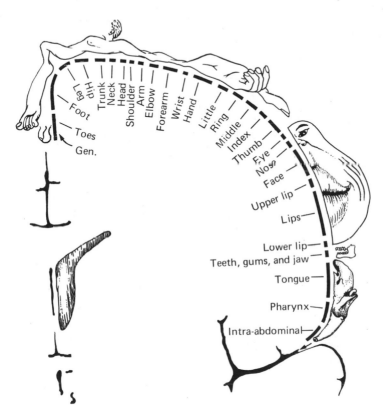

The nerve cells in the brain consist of *cell bodies* each having a long thin process – or *axon* – conducting impulses from the cell. The axons may be very long, sometimes extending from the brain down the spinal cord. The cell bodies also have many finer and shorter fibres, the *dendrites* which conduct signals to the cell (figure 4.3). The cells, with their interconnecting dendrites and their axons, sometimes seem to be arranged randomly, but in some regions they form well ordered

4.3 A nerve cell. The cell body has a long axon, insulated by its myelin sheath, often sending control signals to muscle. The cell body accepts information from the many fine dendrites, some of which tend to make the cell fire, while others inhibit firing. The system is a simple computer element. The inter-connected elements serve to control activity and handle information for perception.

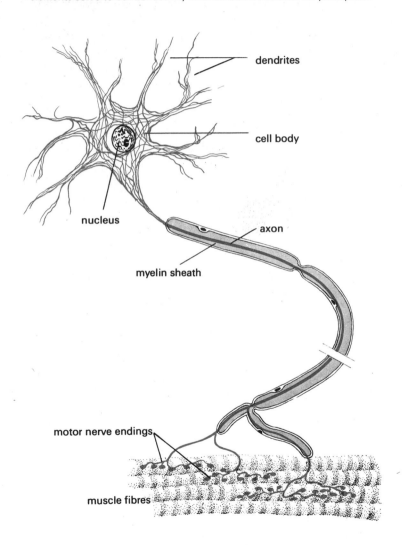

patterns, indicating ordered connections; in the visual cortex they are arranged in layers.

The neural signals are in the form of electrical pulses, which occur when there is an alteration in the ion permeability of the cell membrane (figure 4.4). At rest, the centre of the fibre is negative with respect to the surface; but when a disturbance occurs, as when a retinal receptor is stimulated by light, the centre of the fibre becomes positive, initiating a flow of current which continues down the nerve as a wave. It travels very much more slowly than does electricity along a wire: in large fibres it travels at about 100 metres per second, and in the smallest fibres at less than one metre per second. The thick high-speed fibres have a special fatty coating – the *myelin sheath* – which insulates the fibres from their neighbours and also serves to increase the rate of conduction of the action potential. The low rate of travel – first measured by Helmholtz in 1850 – was a great physiological surprise. It only accounts however for part of the human 'reaction-time' of nearly half a second; for there is also delay from the switching time of the synapses of the brain which route sensory and motor signals, to give perception and behaviour.

There are many sophisticated techniques for studying the nervous system. Electrical activity of individual cells or groups of cells may be recorded, and regions may be stimulated electrically to evoke not only responses but even – in patients undergoing brain operations – sensations. The effects of loss of regions of brain may be discovered, resulting behaviour changes being related to the regions of damage. The effects of drugs or chemicals applied directly to the surface of the brain may be investigated; this is becoming an important area of research, both to establish that new drugs do not have unpleasant psychological side effects and as another technique for changing the state of the brain to discover functions of its various regions.

These techniques, together with examination of the way regions are joined by bundles of fibres, have made it clear that different parts of the brain are engaged in very different functions. But when it comes to discovering the processes going on in each region, even the most refined techniques look rather crude.

It may seem that the most direct way to study the brain is to examine its structure, and stimulate it and record from it. But like electronic devices, it is not at all easy to see how it works from its structure; and the results of stimulation, recording, and removal of parts are difficult to interpret in the absence of a general model of how it works. In order to establish the results of stimulating or ablating the brain it is essential to perform associated behaviour experiments. The results of recording from brain cells are also most interesting when there is some related behaviour, or reported experience. This means that animal and human psychology are very important, for it is essential to relate brain activity to behaviour, and this involves specially designed psychological experiments.

The brain is, of course, an immensely complicated arrangement of nerve cells but it is somewhat similar to man-made electronic devices, so general engineering considerations can be helpful. Like a computer, the brain accepts information, and makes decisions according to the available information; but it is not very similar to actual computers designed by engineers, if only because there are already plenty of brains at very reasonable cost (and they are easy to make by a well-proved method) so computers are designed to be different.

It is easier to make a machine to solve mathematical or logical problems – to handle symbols according to rules – than to see. The problem of making machines to recognise patterns has been solved in various ways for restricted ranges of patterns, but so far there is no neat solution, and no machine comes anywhere near the human perceptual system in range or speed. It is partly for this reason that detailed study of human perception is important. Finding out what we can of human perception is important not only for understanding ourselves: it may also suggest ways in which perception can be achieved by machines. This would be useful for many purposes – from reading documents to exploration of space by robots.

One of the difficulties in understanding the function of the brain is that it is like nothing so much as a lump of porridge. With mechanical systems, it is usually possible to make a good guess at function by considering the structure of the parts, and this is true of much of the body. The bones of the limbs are *seen* to be levers. The position of attachment of the muscles clearly determines their function.

Mechanical and optical systems have parts whose shapes are closely related to their function, which makes it possible to deduce, or at least guess, their function from their shape. It was possible for

4.4 Mechanism of electrical conduction in nerve. Hodgkin, Huxley and Katz have discovered that sodium ions pass to the inside of the fibre. converting its standing negative charge to positive. Potassium ions leak out. restoring the resting potential. This can happen up to a thousand times a second transmitting spikes of potential which run along nerve as signals – by which we know the world through perception and command behaviour,

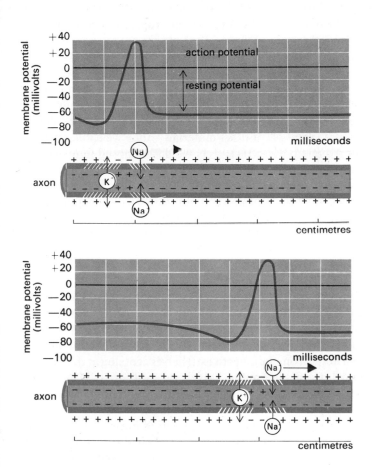

Kepler in the seventeenth century to guess that a structure in the eye (called at that time the 'crystalline') is in fact a lens from its shape. But unfortunately the brain presents a far more difficult problem, if only because the physical arrangement of its parts and their shapes is rather unimportant to their function. When function is not reflected in structure we cannot deduce function by simply looking. It may be necessary to invent an imaginary brain; or invent and perhaps construct a machine, or write a computer program to perform much as the biological system performs. In short, we may have to simulate to explain: but simulations are never complete or perfect. So far we have no machines which approach the brain in thinking or seeing. We have not yet invented anything adequate – so we cannot understand. Nevertheless, some principles may be suggested by engineering considerations. If the brain were strictly unique – impossible to reproduce with symbols or other materials – then we cannot hope to find general explanations. Its secrets would be locked in mysterious properties of protoplasm. But this is hard to believe. If this is not the case we should be able to 'mechanise' brain function; and understand at deep levels the nature of perceptual processes and, perhaps, even consciousness. Meanwhile, we may watch out for developments in machine perception and intelligence to illuminate the brain.

The visual regions

The neural system responsible for vision starts with the retinas. These, as we have seen, are essentially outgrowths of the brain, containing typical brain cells as well as specialised light-sensitive detectors. The retinas are effectively divided down the middle: the optic nerve fibres from the inner (nasal) halves crossing at the chiasma while fibres from the outer halves do not cross. The visual representation corresponds to the brain's representation of touch – in spite of the optical reversal of the eyes (figure 4.5). So touch and vision are closely related. This visual region of brain is known as the *area striata*, from its appearance, the cells being arranged in layers (*frontispiece*).

The brain as a whole is divided down the middle, forming two hemispheres, which are really more or less complete brains, joined by a massive bundle of fibres, the *corpus callosum*, and the smaller

4.5 The optic pathways of the brain. The optic nerve divides at the chiasma, the right half of each retina being represented on the right side of the occipital cortex, the left side on the left half. The lateral geniculate bodies are relay stations between the eyes and the visual cortex.

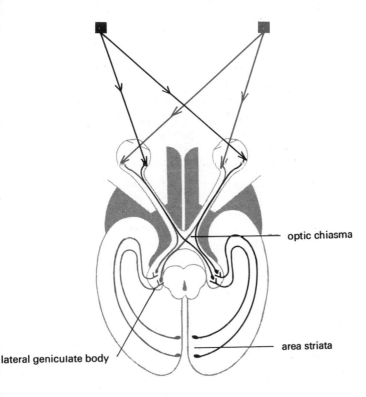

optic chiasma

area striata

lateral geniculate body

optic chiasma. On their way from the chiasma, the optic tracts pass through a relay station in each hemisphere, the *lateral geniculate body*.

The *area striata* is sometimes known as the 'visual projection area'. When a small part is stimulated a human patient reports a flash of light. Upon a slight change of position of the stimulating electrode, a flash is seen in another part of the visual field. It thus seems there is a spatial representation of the retinas upon the visual cortex.

4.6 Hubel and Wiesel's discovery that selected single brain cells (in the cat) fire with movement of the eye in a certain direction. The arrows show various directions of movement of a bar of light presented to the eyes. The electrical record shows that this particular cell fires only for one direction of movement.

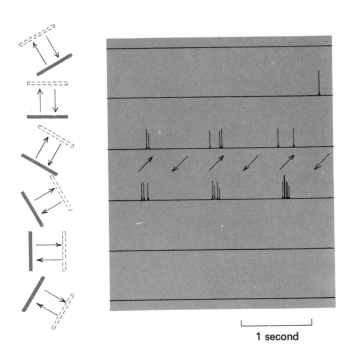

1 second

Stimulation of surrounding regions of the striate area also gives visual sensations, but instead of flashes of light the sensations are more elaborate. Brilliant coloured balloons may be seen floating up in an infinite sky. Further away, stimulation may elicit visual memories, even complete scenes coming vividly before the eyes.

It has recently been found that there is a second projection area – the *superior colliculus* – which gives cruder mapping, and provides command signals to move the eyes. This is evolutionarily more ancient: it seems that the sophisticated feature analysing of the *area*

4.7 Hubel and Wiesel's records from single cells in the visual cortex of the cat. A line (shown on the left) was presented to the cat at various orientations. A single cell in the brain fires only at a certain orientation. This is shown by the spikes of the electrical records.

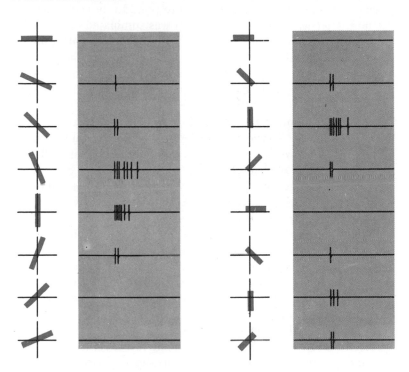

striata has been grafted on to the primitive but still useful – though not conscious – vision of the *colliculus*, as the cortex developed through the evolution of mammals, primates, to man.

Among the most exciting of recent discoveries, is the finding of two American physiologists, David Hubel and Torstin Wiesel, who recorded activity from single cells of the *area striata* of the cat's brain, while presenting its eyes with simple visual shapes. These were generally bars of light, projected by a slide projector on a screen in front of the cat. Hubel and Wiesel found that some cells were only active when the bar of light was presented to the cat at a certain angle. At that particular angle the brain cell would fire, with long bursts of impulses, while at other angles it was 'silent'. Different cells would respond to different angles. Cells deeper in the brain respond to more generalised characteristics, and would respond to these characteristics no matter which part of the retina was stimulated by the light. Other cells responded only to movement, and movement in only a single direction (figure 4.6). These findings are of the greatest importance, for they show that there are specific mechanisms in the brain selecting certain features of objects. Perceptions may be built up from combinations of these selected features.

We do have 'mental pictures', but this should not suggest that there are corresponding electrical pictures in the brain, for things can be represented in symbols – but symbols will generally be very different from the things represented. The notion of brain pictures is conceptually dangerous. It is apt to suggest that these supposed pictures are themselves seen with a kind of inner eye involving another picture, and another eye . . . and so on.

Retinal patterns are represented by coded combinations of activity. The visual cortex is organised not only in the clearly seen layers parallel to its surface, but also in functional 'columns' running through the layers of the striate region. An electrode pushed through the layers, at right angles to them, picks up cells all responding to the same orientation, with more and more general properties as it reaches first the 'simple', then the 'complex' and the 'hypercomplex' cells. If the electrode is inserted about half a millimetre away, then the critical orientation is different. Further away, the signalled orientations are still more different. Each block of (as Colin Blakemore puts it) 'like-minded' orientation detectors is called a 'column'. Various sizes of retinal image features, velocities, and (in monkey and probably man) colours are represented by cells of common orientation down each functional column. Most of the cells are 'binocular' – responding to corresponding points of the two retinas. Gradually the organisation of the 'visual cortex' is becoming clear, though just how this brain activity is related to contours, colours and the shapes of objects as we see them remains mysterious.

5 The eye

Each part of the eye is an extremely specialised structure (figure 5.1). The perfection of the eye as an optical instrument is a token of the importance of vision in the struggle for survival. Not only are the parts of the eye beautifully contrived, but even the tissues are specialised. The cornea is special in having no blood supply: blood vessels are avoided by obtaining nutriment from the aqueous humour. Because of this, the cornea is virtually isolated from the rest of the body. This is fortunate for it makes possible transplants from other individuals in cases of corneal opacity, since antibodies do not reach and destroy it as happens to other alien tissues.

This system of a crucial structure being isolated from the blood stream is not unique to the cornea. The same is true of the lens, and in either case blood vessels would ruin their optical properties. It is also true of a structure in the inner ear, though here the significance is entirely different. In the *cochlea*, where vibrations are converted into neural activity, there is a remarkable structure known as the *organ of Corti*, which consists of rows of very fine hairs joined to nerve cells that are stimulated by the vibration of the hairs. The organ of Corti has no blood supply, but receives its nutriment from the fluid filling the cochlea. If these very sensitive cells were not isolated from the pulse of the arteries, we would be deafened. The extreme sensitivity of the ear is only possible because the crucial parts are isolated from the blood stream, and the same is true of the eye though for a different reason.

The aqueous humour is continually secreted and absorbed, so that it is renewed about once every four hours. 'Spots before the eyes' can be due to floating impurities casting shadows on the retina, seen as hovering in space.

Each eyeball is equipped with six extrinsic muscles, which hold it in position in its orbit, and rotate it to follow moving objects or direct the gaze to chosen features. The eyes work together, so that normally they are directed to the same object, converging for near objects.

5.1 The human eye. The most important optical instrument. Here lies the focusing lens, giving a minute inverted image to an incredibly dense mosaic of light-sensitive receptors, which convert the patterns of light energy into the language the brain can read – chains of electrical impulses.

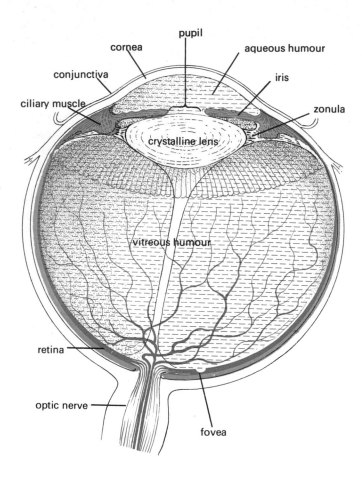

Besides the extrinsic eye muscles, there are also muscles within the eyeball. The iris is an annular muscle, forming the pupil through which light passes to the lens lying immediately behind. This muscle contracts to reduce the aperture of the lens in bright light, and also when the eyes converge to view near objects. Another muscle controls the focusing of the lens. We may look in more detail at the mechanism and function of the *lens* and the *iris*. Both have their surprises.

The crystalline lens. It is often thought that the lens serves to bend the incoming rays of light to form the image. This is rather far from the truth in the case of the human eye, though it is true for fishes. The region where light is bent most in the human eye to form the image is not the lens, but the front surface of the cornea. The reason for this is that the power of a lens to bend light depends on the difference between the refractive index of the surrounding medium and the lens material. The refractive index of the surrounding medium – air – is low, while that of the aqueous humour immediately behind the cornea is nearly as high as that of the lens. In the case of fish, the cornea is immersed in water, and so light is hardly bent at all when it enters the eye. Fish have a very dense rigid lens, which is spherical and moves backwards and forwards within the eyeball to accommodate to distant and near objects. Although the lens is rather unimportant for forming the image in the human eye, it is important for accommodation. This is done not by changing the position of the lens (as in a fish, or a camera) but by changing its shape. The radius of curvature of the lens is reduced for near vision, the lens becoming increasingly powerful and so adding more to the primary bending accomplished by the cornea. The lens is built up of thin layers, like an onion, and is suspended by a membrane, the *zonula*, which holds it under tension. Accommodation works in a most curious manner. For near vision tension is reduced on the zonula, allowing the lens to spring to a more convex form, the tension being released by contraction of the ciliary muscle. It becomes more convex for near vision, by the muscle tightening and not relaxing, which is a surprising system.

The embryological and later developments of the lens are of particular interest, and have dire consequences in middle life. The lens is built up from its centre, cells being added all through life, though growth gradually slows down. The centre is thus the oldest part, and there the cells become more and more separated from the blood system giving oxygen and nutrient, so that they die. When they die they harden, so that the lens becomes too stiff to change its shape for accommodation to different distances.

We see this all too clearly in figure 5.2 which shows how accommodation falls off with age, as the starved cells inside the lens die, and we see through their corpses.

It is possible to see the changes in shape of someone else's lens as he accommodates to different distances. This requires no apparatus beyond a small source of light, such as a flashlight. If the light is held in a suitable position it can of course be seen reflected from the eye, but there is not just one reflection – there are three. The light is reflected not only from the cornea, but also from the front and the back surfaces of the lens. As the lens changes its shape, these images change in size. The front surface gives a large and rather dim image, which is the right way up, while the back surface gives a small bright image which is upside down. The principle can be demonstrated with an ordinary spoon. Reflected from the back convex surface you will see large right-way-up images, but the inside concave surface gives small upside-down images. The size of the images will be different for a large (table) spoon and a small (tea) spoon, corresponding to the curvatures of the lens of the eye for distant and near vision. These images seen in the eye are known as Purkinje images, and are very useful for studying accommodation experimentally.

The iris. This is pigmented, and is found in a wide range of colours. Hence the 'colour of a person's eyes' – which is a matter of some interest to poets, geneticists and lovers, but less so to those concerned with the function of the eye. It hardly matters what colour the iris is, but it must be reasonably opaque so that it is an effective aperture stop for the lens. Eyes where pigment is missing (albinism) are seriously defective.

It is sometimes thought that the changes in pupil size are important in allowing the eye to work efficiently over a wide range of light intensities. This could hardly be its primary function however, for its area only changes over a ratio of about 16:1, while the eye works efficiently over a range of intensity greater than 1,000,000:1. It seems that the pupil contracts to limit the rays of light to the central and optically best part of the lens except when the full aperture is needed for maximum sensitivity. It also closes for near vision, and this increases the depth of field for near objects.

To an engineer, any system which corrects for an external change (in this case light intensity) suggests a 'servo-mechanism'. These are very familiar in the form of the thermostats in central heating, which switch on the heat automatically when the temperature drops below some pre-set value, and then switch it off again when the temperature rises. (An early example of a man-made servo-mechanism is the windmill which aims into the wind and follows its changing direction by means of a fantail sail, which rotates the top of the mill through gearing. A more elaborate example is the automatic pilot which keeps a plane on correct course and height by sensing errors and sending correcting signals to the control surfaces of the machine.)

To go back to the thermostat sensing temperature changes in a central heating system: imagine that the difference between the setting of the lower temperature for switching on the heat is very close to the upper temperature for switching it off. No sooner is it switched on than temperature will rise enough to switch it off again, and so the heating system will be switched on and off rapidly until perhaps something breaks. Now by noting how frequently it is switched on and off, and noting also the amplitude of the temperature variation, an engineer could deduce a great deal about the system. With this in mind, some subtle experiments have been performed on just how the iris servo-control system works.

The iris can be made to go into violent oscillation, by directing a narrow beam of light into the eye, so that it passes the edge of the iris (figure 5.4). When the iris closes a little the beam is partly cut off, and so the retina gets less light. But this gives the iris a signal to open. As soon as it opens the retina gets more light – and then the iris starts to close, until it gets another signal to open. Thus it oscillates in and out. By measuring the frequency and amplitude of oscillation of the iris a good deal can be learned about the neural servo-system controlling it.

The pupil. This is not, of course, a structure. It is the hole formed by the iris through which light passes to the lens and on to the retina as

5.2 (Top) Loss of accommodation of the lens of the eye with ageing. The lens gradually becomes rigid and cannot change its form. Bifocal spectacles serve to give effective change of focus when accommodation is lost.
5.3 (Bottom) Eye **a** cannot see into eye **b**. One's own eye always gets in the way, preventing light reaching the only part of the retina which could form an image.

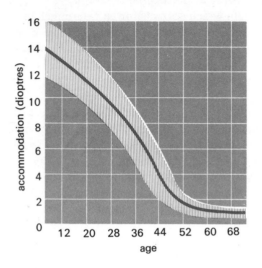

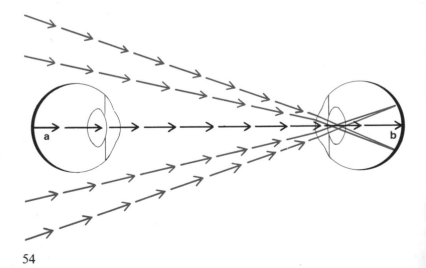

5.4 Making the iris oscillate with a beam of light. When the iris opens a little, more light reaches the retina, which then signals the retina to close. But when it closes less light reaches the retina, which signals the iris to open. Thus it oscillates. From the frequency and amplitude of oscillation the iris control system can be described in terms of servo-theory.

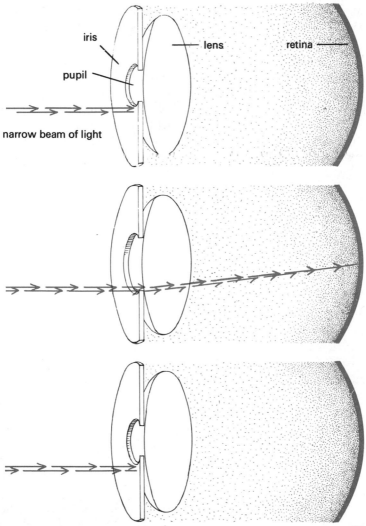

iris

pupil

lens

retina

narrow beam of light

5.5 (right) How eyes look when we can see into them.
This photograph is taken with an ophthalmoscope.
It shows the yellow spot over the fovea, the retinal
blood vessels through which we see the world, and the blind
region where the vessels and nerves leave the eyeball.

an image. Although the human pupil is circular, there is a great variety of shapes, the circular form being rather unusual in nature.

The pupil looks black, and we cannot see through it into another person's eye. This requires some explanation, for the retina is not black, but pink: indeed it is a most curious fact that although we see out of our pupils, we cannot see into someone else's! The reason is that the lens in the other person's eye focuses light from any given position on to a certain region of the retina, so the observing eye always gets in the way of the light which would shine on to the part of the retina he should be seeing (figure 5.3). Helmholtz (anticipated by Charles Babbage) invented a device for looking into another person's eye, the trick being to direct a beam of light along the path the observer is looking (figure 5.6). With this device the pupil no longer looks black, and the detailed structure of the living retina may be seen, inside the eye, the blood vessels on its surface appearing as a great red tree of many branches.

Eye movements

Each eye is moved by six muscles (figure 5.7). The remarkable arrangement of the *superior oblique* can be seen in the illustration. The tendon passes through a 'pulley' in the skull, in front of the suspension of the eyeball. The eyes are in continuous movement, and they move in various ways. When the eyes are moved around, searching for an object, they move quite differently from the way they move when a moving object is being followed with the eyes. When searching, they move in a series of small rapid jerks, but when following they move smoothly. The jerks are known as *saccades* (after an old French word meaning 'the flick of a sail') Apart from these two main types of movement, there is also a continuous small high-frequency tremor.

Eye movements can be recorded in various ways: they can be

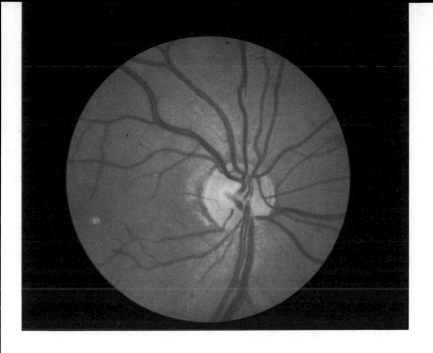

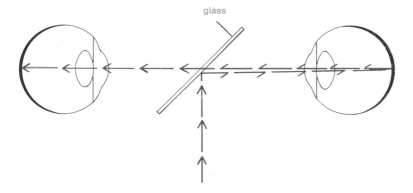

5.6 The ophthalmoscope, invented by Charles Babbage and Helmholtz. Light reaches the observed eye by reflection *from* a half-silvered mirror, *through* which the observer sees inside the illuminated eye. (In practice he may look above the illuminated ray to avoid losses of the half-silvered mirror).

5.7 The muscles which move the eye. The eyeball is maintained in position in the orbit by six muscles, which move it to direct the gaze to any position and give convergence of the two eyes for depth perception. They are under continuous tension and form a delicately balanced system, which when upset can give illusions of movement.

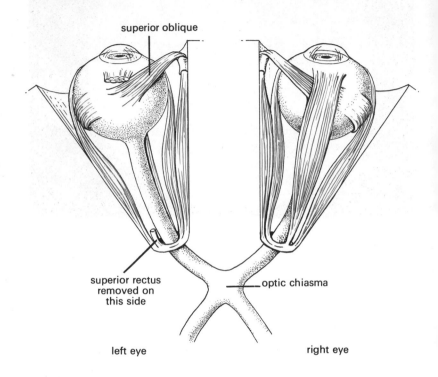

superior oblique

superior rectus
removed on
this side

optic chiasma

left eye   right eye

filmed with a cine camera, detected by small voltage changes around the eyes, or most accurately by attaching a mirror to a contact lens placed on the cornea, when a beam of light may be reflected off the mirror and photographed on continuously moving film.

It turns out that the saccadic movements of the eyes are essential to vision. It is possible to fix the image on the retina so that wherever the eye moves, the images move with it and so remain fixed on the retina. When the image is optically stabilised (figure 5.8) vision fades after a few seconds, and so it seems that part of the function of eye

5.8 A simple way of optically stabilising the retinal image. The object (a small photographic transparency), is carried on the eye on a contact lens, and moves exactly with it. After a few seconds the eye becomes blind to the stabilised image, some parts fading before others. This method was devised by R. Pritchard.

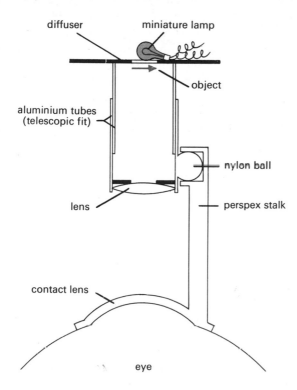

diffuser

miniature lamp

object

aluminium tubes (telescopic fit)

nylon ball

lens

perspex stalk

contact lens

eye

movements is to sweep the image over the *receptors* so that they do not adapt and so cease to signal to the brain the presence of the image in the eye. But there is a curious problem: when we look at a sheet of white paper, the edges of the image of the paper will move around on the retina, and so stimulation will be renewed; but consider now the centre of the image. Here the small movements of the eyes can have no effect, for a region of given brightness is substituted for another region of exactly the same brightness, and so no change in stimulation takes place with the small eye movements. Yet the middle of the paper does not fade away. This suggests that borders and outlines are very important in perception. Large areas of constant intensity provide no information. They seem to be 'inferred' from the signalled borders: the central visual system makes up the missing signals.

Blinking is often assumed to be a reflex, initiated by the cornea becoming dry. But for normal blinking this is not so; though blinking can be initiated by irritation of the cornea, or by sudden changes in illumination. Normal blinking occurs with no external stimulus: it is initiated by signals from the brain. The frequency of blinking increases under stress, and with expectation of a difficult task. It falls below average during periods of concentrated mental activity. Blink rate can even be used as an index of attention and concentration on tasks of various kinds and difficulties.

The retina
The name retina is derived from an early word meaning 'net' or 'cobweb tunic', from the appearance of its blood vessels.

The retina is a thin sheet of interconnected nerve cells, including the light-sensitive rod and cone cells which convert light into electrical pulses – the language of the nervous system. It was not always obvious that the retina is the first stage of visual sensation. The Greeks thought of the retina as providing nutrient to the vitreous. The source of sensation was supposed by Galen (c. 130–201 A.D.) and by much later writers, to be the crystalline lens. The Arabs of the 'dark' ages – who did much to develop optics – thought of the retina as conducting the vital spirit, the 'pneuma'.

It was the astronomer Kepler who, in 1604, first realised the true function of the retina – that it is the screen on which an image from the lens is formed. This hypothesis was tested experimentally by Scheiner, in 1625. He cut away the outer coating (the *sclera* and the *choriod*) from the back of an ox's eye, leaving the retina revealed as a semi-transparent film – to see an upside-down image on the retina of the ox's eye.

The discovery of the photoreceptors had to wait upon the development of the microscope. It was not until about 1835 that they were first described, and then none too accurately, by Treviranus. It seems that his observation was biased by what he expected to see, for he reported that the photoreceptors face the light. Strangely, they do not: in mammals and in nearly all vertebrates – though not in cephalopods – the receptors are placed at the back of the retina, behind the blood vessels. This means that light has to go through the web of blood vessels and the fine network of nerve fibres – including three layers of cell bodies and a host of supporting cells – before it reaches the receptors. Optically, the retina is inside out, like a camera film put in the wrong way round (figure 5.9). Given the original 'mistake' however (which seems to result from the embryological development of the retina from the surface of the brain), the situation is largely saved by the nerve fibres from the periphery of the retina skirting around and avoiding the crucial central region giving best vision.

The retina has been described as 'an outgrowth of the brain'. It is a specialised part of the surface of the brain which has budded out and become sensitive to light, while it retains typical brain cells functionally between the receptors and the optic nerve (but situated in the front layers of the retina) which greatly modify the electrical activity from the receptors themselves. Some of the data processing for perception takes place in the eye, which is thus an integral part of the brain.

There are two kinds of light-receptor cells – the *rods* and the *cones* – named after their appearance as viewed with a microscope. In the peripheral regions of the retina they are clearly distinguishable, but in the central region – the *fovea* – the receptors are packed exceedingly close together, and look like the rods.

The cones function in daylight conditions, and give colour vision. The rods function under low illumination, and give vision only of shades of grey. Daylight vision, using the cones of the retina, is referred to as *photopic* while the grey world given by the rods in dim light is called *scotopic*.

It might be asked how we know that only the cones mediate colour vision. This is deduced partly from studies of various animal eyes, by relating retinal structure to their ability to discriminate colours as determined by behaviour experiments; and also from the finding that in the human retina there are very few cones near the edge of the retina, where there is no colour vision. It is interesting that although the central foveal region, packed tightly with functional cones, gives

5.9 The retina. Light travels through the layers of blood vessels, nerve fibres and supporting cells to the sensitive receptors ('rods' and 'cones'). These lie at the back of the retina, which is thus functionally inside-out. The optic nerve is not, in vertebrate eyes, joined directly to the receptors, but is connected via three layers of cells, which form part of the brain externalised in the eyeball.

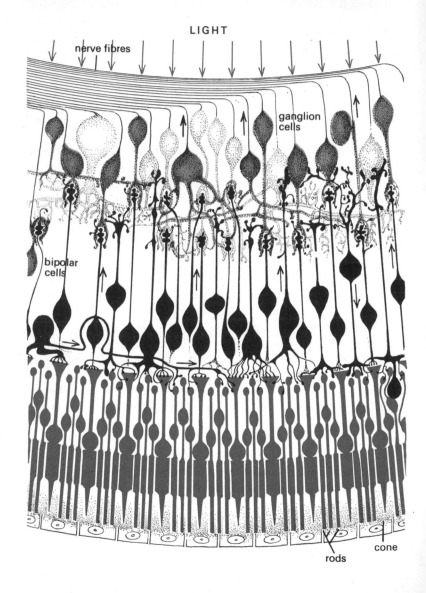

LIGHT

nerve fibres

ganglion cells

bipolar cells

rods

cone

the best visual detail and colour, it is less sensitive than the more primitive rod-regions of the retina. (Astronomers 'look off' the fovea when they wish to detect very faint stars so that the image falls on a region of the retina rich in sensitive rods).

It might be said that by moving from the centre of the human retina to its periphery we travel back in evolutionary time; from the most highly organised structure to a primitive eye, which does little more than detect movements of shadows. The very edge of the human retina does not even give a sensation when stimulated by movement. It gives primitive unconscious vision; and directs the highly developed foveal region to where it is likely to be needed for its high acuity.

The size of the receptors and their density become important when we consider the ability of the eye to distinguish fine detail. We shall quote directly from Polyak's great book *The Retina*:

The central territory where the cones are almost uniformly thick measures approximately $100\,\mu$ (microns, or millionths of a metre) across, corresponding to 20', or one-third of a degree of arc. It contains approximately fifty cones in a line. This area seems to be not exactly circular but elliptical, with the long axis horizontal, and may contain altogether 2,000 cones . . . the size of each of the 2,000 receptor-conductor units measures, on the average, 24" of arc. The size of the units even in this territory varies, however, the central most measuring scarcely more than 20" of arc or even less. Of these – the most reduced cones, and therefore the smallest functional receptor units – there are only a few, perhaps not more than one or two dozen. The size of the units given includes the intervening insulating sheaths separating the adjoining cones from one another.

It is worth trying to imagine the size of the receptors. The smallest is one micron, only about two wavelengths of red light in size. One could not ask for much better than that. Even so, the visual acuity of the hawk is four times better than that of man.

The number of cones is about the same as the population of Greater New York. If the whole population of the United States of America were made to stand on a postage stamp, they would represent the rods on a single retina. As for the cells of the brain – if people were scaled down to their size, we could hold the population of the earth in our cupped hands; but there would not be enough people to make one brain.

The photopigments of the retina are bleached by bright light; it is this bleaching which – by some still mysterious process – stimulates the nerves, and it takes some time for the photochemical to return to normal. The retinal chemical cycle involved is now understood, primarily through the work of George Wald. While a region of photopigment is bleached, this region of the retina is less sensitive than the surrounding regions, and this gives rise to *after-images*. When the eye has been adapted to a bright light (e.g. a lamp bulb viewed with the eye held steady, or better a photographic flash) a dark shape, of the same form as the adapting light, is seen hovering in space. It is dark when seen against a lighted surface, but for the first few seconds it will look bright, especially when viewed in darkness. This is called a 'positive' after-image, and represents continuing firing of the retina and optic nerve after the stimulation. When dark, it is called a 'negative' after-image, and represents the relatively reduced sensitivity of the stimulated part of the retina due partly to bleaching of the photopigment.

Two eyes

Many of the organs of the body are duplicated, but the eyes and also the ears are unusual in working in close co-operation: for they share and compare information, so that together they perform feats impossible for the single eye or ear.

The images in the eyes lie on the curved surfaces of the retinas, but it is not misleading to call them two-dimensional. A remarkable thing about the visual system is its ability to synthesise the two somewhat different images into a single perception of solid objects lying in three-dimensional space.

In man the eyes face forwards, and share the same field of view; but this is rare among vertebrates, for generally the eyes are at the sides of the head and aim outwards in opposed directions. The gradual change from sideways to frontal looking eyes came about as precise judgment of distance became important, when mammals developed front limbs capable of holding and handling objects, and catching the branches of trees. For animals who live in forests and travel by leaping from branch to branch, rapid and precise judgment of distance of nearby objects is essential; and the use of two eyes, co-

5.10 The eyes converge on an object we examine, the images being brought to the foveas. In **a** we see the eyes converged to a near object, in **b** to one more distant. The angle of convergence is signalled to the brain as information of distance – serving as a range-finder.

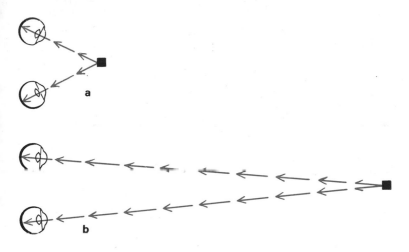

operating to give stereo vision, is highly developed. Animals such as the cat have frontal eyes which function together, but the density of the receptors is nearly constant over the retina. There is no fovea unless precise depth perception is really important, as in birds and in the tree-living apes, when we find developed foveas and precise control of eye movements. Stereo vision for movement is also provided by the paired compound eyes of insects, and is highly developed in insects such as the dragon fly which catches its prey at high speeds on the wing. The compound eyes are fixed in the head, and the mechanism of their stereo vision is simpler than in apes and man where foveas are brought to bear on objects at different distances by convergence of the eyes.

Convergence, or range-finder, depth perception   Figure 5.10 shows how the eyes pivot inwards for viewing near objects: distance is signalled to the brain by this angle of convergence. This, however, is by no means the whole story.

A simple experiment shows that the convergence angle is indeed

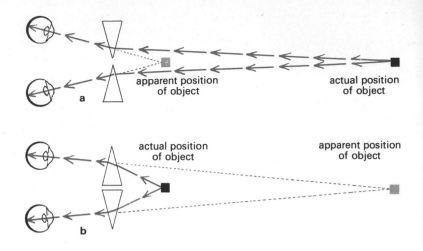

apparent position
of object

actual position
of object

a

actual position
of object

apparent position
of object

b

used to signal distance. Figure 5.11a shows what happens if a pair of prisms of suitable angle are introduced to bend the light entering the eyes, so that they have to converge to bring distant objects on to the centre of the foveas. If the prisms are placed to decrease the angle of convergence (figure 5.11b) objects will appear nearer and larger; with prisms arranged to increase convergence objects appear further and smaller.* Depth perception is given in part by the angle of convergence of the eyes indicating distance, just as in a range-finder.

Now there is a serious limitation to range-finders: they can only indicate the distance of one object at a time; namely, the object whose images are fused by the appropriate convergence angle. To find the distances of many objects at the same time it is necessary to adopt a very different system. The visual apparatus has developed such a system, but it involves some elaborate computation for the brain.

Disparity depth perception The eyes are horizontally separated (by about 6.3 cm) and so receive somewhat different views. This can be seen quite clearly if first one eye then the other is held open. Any near object will appear to shift sideways in relation to more distant objects,

* The size changes however are complicated, because the total depth scale is shrunk or expanded by convergence. Cf. page 73.

5.11 The angle of convergence for a given distance can be changed by interposing prisms. **a** shows increased, and **b** decreased, convergence. The effect is to change the apparent size and distance of objects viewed through the prisms. The change is not optical, but due to rescaling by the brain, its range-finder giving it the wrong information. This is a useful experimental trick for establishing the importance of convergence on perception of size and distance.

and to rotate when each eye receives its view. This slight difference between the images is known as 'disparity'. It gives perception of depth by *stereoscopic vision*. This is employed in the stereoscope.

The stereoscope (invented by Sir Charles Wheatstone in about 1832) is a simple instrument for presenting any two pictures separately to the two eyes. Normally these pictures are stereo pairs, made with a pair of cameras separated by the distance between the eyes, to give the disparity which the brain uses to give stereo depth vision.

Stereo pictures may be presented reversed – the right eye receiving the left eye's picture and *vice versa* – and then we may get reversal of perceived depth. Depth reversal always occurs with this pseudoscopic vision (as it is called) except when the reversed depth would be highly unlikely. People's faces will not reverse in depth. Hollow faces are too improbable to be accepted. So stereopsis is but a cue to depth, which may be rejected (figure 5.12).

Stereo vision is only one of many ways in which we see depth, and it only functions for comparatively near objects, after which the difference between the images becomes too small: we are effectively one-eyed for distances greater than perhaps 100 metres.

The brain must 'know' which eye is which, for otherwise depth perception would be ambiguous. Also, reversal of the pictures in a stereoscope (or a pseudoscope) would have no effect. But oddly enough, when the light is cut off to one eye it is virtually impossible to *say* which eye is doing the seeing. Although the eyes are fairly well identified for the depth mechanism, this information is not available to consciousness.

If the pictures presented to the two eyes are very different (or if the difference between the viewing positions of an object is so great that the corresponding features fall outside the range for fusion) a curious and highly distinctive effect occurs. Parts of each eye's picture are successively combined and rejected. This is known as 'retinal rivalry'.

5.12 Switching the eyes with mirrors. (Top) A *pseudoscope* – gives reversed depth, but only when depth is somewhat ambiguous. (Centre) A *telestereoscope* – effectively increases the separation of the eyes. (Bottom) An *iconoscope* – reduces the effective distance apart of the eyes. These arrangements are all useful for studying convergence and disparity in depth perception.

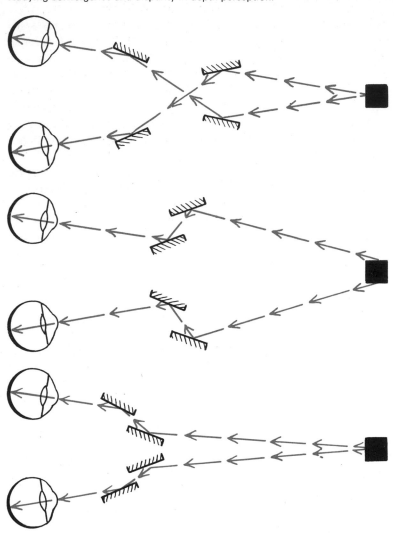

Rivalry also generally occurs if different colours are presented to two eyes, though fusion into mixture colours is possible, especially when contours are fused by the eyes.

A great deal has been discovered over the last few years of how the brain combines the images of the two eyes, using the minute 'disparity' differences to compute depth. We now know that most cells in the visual cortex respond to stimulation from corresponding points of the two retinas. These are called 'binocular cells'. Presumably corresponding retinal points, as related by disparity to distance, are indicated by the firing of these binocular cells.

Until recently it has been assumed that stereoscopic vision always functions by binocular comparison between the edges of objects. In what has turned out to be an unusually important technique, Bela Julesz, of the Bell Telephone Laboratories in America, has shown that lines or borders are not needed. Julesz generates pairs of random dot patterns with a computer, so arranged that for each dot in the pattern shown to one eye there is a corresponding dot for the other eye. When groups of corresponding dots are displaced horizontally, the displaced dots are fused with their corresponding dots of the other eye's field – and depth is seen. Regions of the random patterns stand out (or sink behind) the rest, and may be given complex three dimensional shapes. This shows that cross correlation of points even where there are no contours can give stereoscopic vision. This fits the 'binocular cells' discovered by micro-electrode recording in the brain. A pair of Julesz stereograms is shown in figure 5. 13. These need to be fused with a stereoscope. Many beautiful examples will be found in Julesz' book *The Cyclopean Eye* (1971).

Is this all there is to stereo vision? As usual – there is more to it! In the first place it is not clear why some dots but not others are accepted as 'corresponding'. This needs some kind of global decision by the visual system. Further, we see illusory contours joining groups of dots standing out in depth from the random background. This is some kind of contour creation. We can also create illusory contours for each eye with patterns such as figure 5.14. Can we produce *stereopsis* with such illusory contours? The answer is, Yes. Figure 5.15 shows a pair of illusory contours having disparity, which when presented one to each eye in a stereoscope give stereoscopic depth – though there is no stimulation of binocular cells. This suggests that

5.13 These random patterns were computer-generated, with a lateral shift of each point in a central region of one of the patterns. When seen, one with each eye (using a stereoscope), this central region appears above the rest of the pattern – showing that the eyes perform a cross-correlation between the pair of patterns, to convert disparity (horizontal displacement) into perception of depth. Stereoscopic depth does not therefore depend upon contours, as had been thought before Julesz's experiments.

stereopsis can work not only from the physical points of stimulation at the sense organs – but also from *inferred* contours. On this (cognitive) explanation, illusory contours are *postulated* by the brain when gaps are unlikely – and are very probably due to masking by some nearer object hiding part of the figure. A nearer masking object is then postulated, from the evidence of the unlikely gaps. If this account is correct, we see that *absence* of stimulation can serve as *data* for perception. Similarly, the illusory letters seen in figure 10.17 seem to be postulated, or guessed, from the shadows. So even simple

perceptions, such as contours, may have subtle and 'cognitive' origins.

This account of illusory contours is not universally accepted. Some authorities suppose that these are contrast effects (cf. figure 6.1) or that they are given by neural interactive effects such as lateral inhibition. The issue is whether these illusory contours are produced indirectly, by *data* (surprising absence of stimulation) or more directly by (disturbed) *signals*, due to interactive effects early in the visual system. The first kind of theory is cognitive and the second is purely and simply physiological.

5.14 These broken ray figures (and there are many other examples) produce illusory contours, and whiter-than-white or blacker-than-black regions. It seems that they are 'postulated' as masking objects lying in front of improbable gaps. If so, to seek simple relations between physiological activity of, for example 'feature detectors' and experience of even simple features such as contours, is to be optimistic.

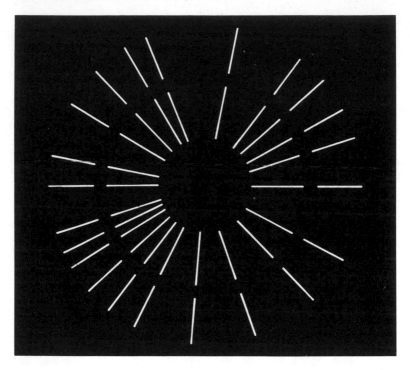

The issue is fundamental. On the cognitive account – that these illusory contours are inferred – we can hardly expect to find simple relations between neural activity (such as from the 'feature detectors' [pages 46–8]) and even elementary perceptions, such as contours.

Some people do not get depth from the Julesz dot figures, though they do from ordinary line or picture stereograms. Also, for normal observers, if there is no *brightness contrast*, but only colour contrast – then the dot figures do not give depth – though depth is still seen from lines. So there may be two brain mechanisms for stereo depth.

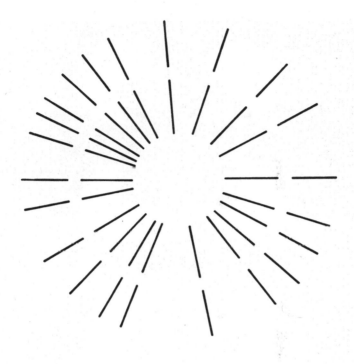

We end with a final feature of stereo depth perception. There is a clear linkage between two of the depth signalling mechanisms we have described – (1) the convergence of the eyes serving as a *range finder*, and (2) the difference between the two images giving *disparity*. The angle of convergence adjusts the scale of the disparity system. When the eyes fixate a *distant* object, disparities between the images are accepted as representing *greater differences in depth* than when the eyes are converged for near vision.

If this did not occur, distant objects would look closer together in

5.15 Shows a pair of figures which produce curved illusory contours. When presented to each eye — we observe stereoscopic depth from disparity of illusory contours (to be viewed horizontally with a stereoscope).

depth than near objects of the same depth separation, for the disparity is greater the nearer the objects. The linked mechanism compensating for this geometrical fact may be seen at work quite easily – by upsetting convergence while keeping the disparity the same. If the eyes are made to converge to infinity (with prisms) though near objects are being observed, they appear stretched out in depth. So we can see our convergence-disparity compensation system at work.

Is contour perception innate or learned? Recent experiments have attempted to decide whether the orientation feature detectors (as they are often though probably misleadingly called) are given innately or whether early experience affects them. Kittens have been reared in environments of vertical stripes, and have been tested for vision of vertical and for other orientations of stripes – and for corresponding feature detectors. It has been found, by Colin Blakemore and others, that kittens living in a world of only vertical stripes appear to be blind to horizontal stripes, and they lack horizontal feature detectors. Similarly, kittens denied horizontal stripes do not have the usual vertical feature detectors. Although some investigators have failed to repeat this result, it is now generally accepted. It suggests that not all the basic feature detectors are laid down at birth, but are developed by visual stimulation. Alternatively, such innate neural mechanisms may degenerate with lack of stimulation. In either case, the early visual environment of babies may be highly important for adult vision – so nursery wallpaper should be considered!

There are both practical and theoretical implications to this kind of research into the origins of neural connections and properties of brain mechanisms. How much detailed 'wiring' is laid down by genetic instructions? How flexible, how adaptable, is the nervous system? Can lost time be made up later in life? These are pioneer experiments relating physiology to knowledge gained from experience.

6.1 Simultaneous contrast. The part of the grey ring seen against the black appears somewhat lighter than the rest, seen against white. This effect is enhanced if a fine thread is placed across the ring along the black-white junction.

6 Seeing brightness

There is supposed to be a primitive tribe of cattle breeders who have
no word for green in their language, but have six words for different
shades of red – specialists in all fields adopt special meanings for their
own use. Before embarking on a discussion of brightness and colour,
we should stop for a moment to sharpen some words – as a carpenter
might stop to sharpen his chisels before attempting delicate work.

We speak of *intensity* of light entering the eye, giving rise to
brightness. Intensity is the physical energy of the light, which may be
measured by various kinds of photometer, including the familiar
photographer's exposure meter. Brightness is an experience. We
believe we know what another person means when he says 'what a
bright day!' He means not only that he could take photographs with a
slow film in his camera, but also that he experiences a dazzling
sensation. This sensation is roughly, but only roughly, related to the
intensity of the light entering the eyes.

When talking technically about colour vision, we do not generally
talk of 'colours', but rather of 'hues'. This is simply to avoid the
difficulty that 'colours' are apt to mean sensations to which we can
give a specific name, such as 'red' or 'blue'. We thus speak technically
of 'spectral hues', rather than 'spectral colours' but this is not always
necessary. The intensity-brightness distinction is more important.

Another important distinction to be made is *colour as a sensation*
and *colour as a wavelength* (or set of wavelengths) of the light entering
the eye. Strictly speaking, light itself is not coloured: it gives rise to
sensations of brightness and colour, but only in conjunction with a
suitable eye and nervous system. The technical language is somewhat
confused on this matter: we do speak sometimes of 'coloured light',
such as 'yellow light', but this is loose. It should be taken to mean:
light which generally gives rise to a sensation described by most
people as 'yellow'.

Without attempting to explain how physical intensities and
wavelengths of radiation give rise to different sensations (and

ultimately we do not know the answer) we should realise quite clearly that without life there would be no brightness and no colour. Before life came, all was silent though the mountains toppled.

The simplest of the visual sensations is brightness. It is impossible to describe the sensation. A blind man knows nothing of it, and yet to the rest of us reality is made up of brightness and of colour. The opposed sensation of blackness is as powerful – we speak of a 'solid wall of blackness pressing in on us' – but to the blind this also means nothing. The sensation given to us by absence of light is blackness; but to the blind it, like light, is nothing. We come nearest to picturing the world of the blind, who have no brightness and no black, by thinking of the region behind our heads. We do not experience blackness behind us: we experience nothing, and this is very different from blackness.

Brightness is not just a simple matter of the intensity of light striking the retina. The brightness given by a given intensity depends upon the state of adaptation of the eye, and also upon various complicated conditions determining the contrast of objects or of patches of light. In other words, brightness is a function not only of the intensity of light falling on a given region of the retina at a certain time; but also of the intensity of the light that the retina has been subject to in the recent past, and of the intensities of light falling on other regions of the retina.

Dark-light adaptation

If the eyes are kept in a low light level for some time they grow more sensitive, and a given light will look brighter. This 'dark adaptation' is rapid for the first few seconds, then slows down. The rod and cone receptor cells adapt at different rates: cone-adaptation is completed in about seven minutes, while rod-adaptation continues for an hour or more. This is shown in figure 6.2, where it will be seen that there are two adaptation curves – one for the rods, the other for the cones. It is as if we have not one but two retinas, lying inter-mingled in the eye.

The mechanisms of dark adaptation are beginning to be understood in detail, largely through the ingenious and technically brilliant experiments of the British physiologist W. A. H. Rushton. It was suggested many years ago that adaptation is due to regeneration of

6.2 Increase in sensitivity of the eye in the dark, known as *dark adaptation*. The *red* curve shows how the cone cells adapt, while the *black* curve shows rod adaptation, which is slower and proceeds to greater sensitivity. In dim light only the rods are functional, while they are probably inhibited in brighter light used by the active cones.

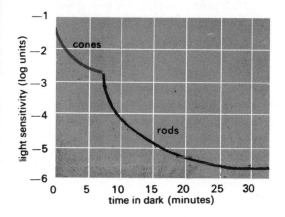

the visual pigments of the eye bleached by light – this bleaching in some unknown way stimulating the receptors to give the electrical signals to the optic nerve. The photochemical rhodopsin was extracted from the frog's eye, and its density to light measured during bleaching and regeneration, and compared with human dark adaptation curves. The two curves are shown together in figure 6.4, and indeed they do correspond very closely, suggesting a strong connection between the photochemistry of rhodopsin and the changing sensitivity of the rod eye. It would also seem that brightness must be related to the amount of photochemical present to be bleached. What Rushton has done is to measure the density of the photochemical in the living eye, during adaptation to darkness or to any coloured light. The technique is, essentially, to shine a brief flash of light into the eye and to measure the amount of light reflected from it, with a very sensitive photocell. At first it seemed impossible to do this for the human eye because so little light remains to be reflected after the almost complete absorption by the photochemicals and the black pigment lying behind the receptors. So a cat's eye was used, the reflecting layer at the back, the tapetum, serving as a mirror to reflect light to the photocell. The method worked with the cat's eye; and Rushton then succeeded in making it sufficiently sensitive to detect and measure the very feeble light reflected out from the human eye. He found that there is bleaching of the photochemicals with adaptation, though this was less than expected. He then detected the three colour-sensitive pigments in this way, a result which has been confirmed by microscopic examination of individual cone cells.

Contrast

Another factor which affects brightness is the intensity of surrounding areas. A given region generally looks brighter if its surroundings are dark, and a given colour looks more intense if it is surrounded by its complementary colour. This is no doubt related to the cross-connections between the receptors. Contrast enhancement seems to be tied up with the general importance of borders in perception. It seems that it is primarily the existence of borders which are signalled to the brain: regions of constant intensity not requiring much information. The visual system extrapolates between borders, which no doubt saves a lot of information-handling by the peripheral parts of the system, though at the cost of some complexity further up in the visual channel. This probably starts with retinal *lateral inhibition*. Although the phenomena of contrast and enhancement of borders are no doubt mainly due to retinal mechanisms, there do seem to be more central contributions. This is brought out in figure 6.1 which shows quite marked contrast; the even grey ring appears lighter where it lies against the dark background than where it lies against the white. But this effect is considerably more marked when a fine thread is placed across the ring continuing the division of the background; the contrast is greater when the figure is interpreted as two separate halves than when it is regarded as all one figure. This suggests that central brain factors play a part.

Something of the subtlety of the human brightness system is shown by Fechner's Paradox. This is as follows. Present the eye with a small, fairly bright source: it will look a certain brightness, and the pupil will close to a certain size when the light is switched on. Now add a second, dimmer light. This is placed some way from the first, so that a different region of retina is stimulated. What happens? Although the total intensity has increased with the addition of the second light, the pupil does not close further, as one might expect: rather it opens, to correspond to an intensity between the first and the second light. It is evidently set not by the *total*, but by the *average* illumination.

Try shutting one eye, and noting any change in brightness. There is practically no difference whether one or two eyes receive the light. This, however, is not so when small dim lights are viewed in surrounding darkness: they *do* look considerably brighter with two eyes than with one. This phenomenon is not understood.

Brightness is a function of colour. If we shine lights of different colours but the same intensity into the eyes, the colours at the middle of the spectrum will look brighter than those at the ends. This is shown in figure 6.5, the curve being known as the *spectral luminosity curve*. This is of some practical importance, for if a distress signal light is to be clearly visible, it should be of a colour to which the eye is maximally sensitive – in the middle of the spectrum. The matter is complicated by the fact that the sensitivity curves for rods and cones are somewhat different. They are similar in general shape, but the cones are most sensitive to yellow, while the rods are most sensitive to green. The change with increasing intensity is known as the Purkinje Shift.

The luminosity curve tells us nothing much about colour vision. It is sensitivity to light plotted against wavelength of light, but with no reference to the colours seen at each wavelength. Animals without colour vision show a similar luminosity curve.

It seems that although there are photochemical changes associated with adaptation to light, there are several additional mechanisms at work, these being not photochemical but neural. In particular, as the eye becomes dark adapted, it trades its acuity in space and in time for increase in sensitivity. With decrease of intensity, and the compensating dark adaptation, ability to make out fine detail is lost. This is no simple matter, but it is in part due to the retina integrating energy over a greater area – over a greater number of receptors. There is also an increase in the time over which photic energy is integrated as the eye adapts to dim light, much as photographers use longer exposures to compensate for low light levels.

The trading of temporal discrimination for sensitivity with dark

6.3 The Pulfrich Pendulum. A pendulum swinging in a straight arc across the line of sight is viewed with a dark glass over one eye, both eyes being open. It appears to swing in an ellipse. This is due to the signals from the eye which is dark-adapted by the filter being delayed. The bob's increasing velocity towards the centre of its swing gives increasing *signalled* disparity: accepted as stereo depth signals corresponding to an ellipse.

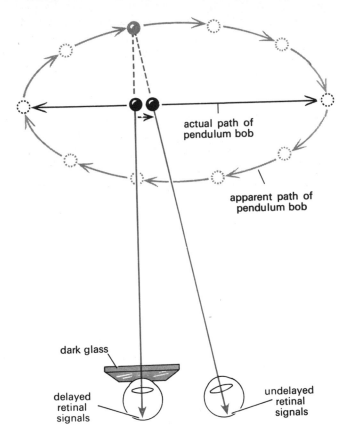

actual path of
pendulum bob

apparent path of
pendulum bob

dark glass

delayed
retinal
signals

undelayed
retinal
signals

adaptation is elegantly, if somewhat indirectly, observed in a curious and dramatic phenomenon known as the *Pulfrich Pendulum effect*. Not least remarkable about this effect is its discovery, for the phenomenon cannot be seen without two functional eyes, and yet its discoverer was blind in one eye! The experiment is worth trying out. Take a length of string, and a weight for a bob to make a pendulum about one metre long. Swing the pendulum in a straight arc normal to the line of sight. View the oscillating bob with both eyes, but cover one with a dark, though not opaque, glass such as one filter of a pair of sun glasses.

It will be found that the bob does not appear to swing in a straight arc, but to *describe an ellipse*. The ellipse may be extremely eccentric – indeed the longer axis can lie along the line of sight, though the bob is actually swinging straight across the line of sight.

Now what causes this strange effect? By reducing the light, the dark glass delays signals from this eye. The receptors take longer to respond, and the dark adaptation produces a delay in the message reaching the brain from this eye. The delay causes the affected eye to see the bob slightly in the past, and as the bob speeds up in the middle of its swing, this delay becomes more important, for the eye with the filter sees it in a position further and further behind the position signalled to the brain by the unaffected eye. This gives an effective horizontal shift of the moving image – as signalled – generating stereo depth. For the brain it is as though the bob is swinging elliptically. This is shown in figure 6.3. It seems that increased delay with dark-adaptation is associated with increase in temporal integrating time: as when a photographer uses a longer exposure in dim light. We see this also, and more directly, from the 'comet's tail' following a moving firework seen at night, as dark adaptation increases the effective exposure-time of the eye to increase its sensitivity.

Both the increase in the delay of messages from the retina to the brain, and the increase in the integrating time which this allows, have some practical significance. The retinal delay produces a lengthening of reaction-time in drivers in dim light, and the increased integrating time makes precise localisation of moving objects more difficult. Games cannot be played so well: the umpire calls 'Cease play for poor light' long before the spectators think it right to bow before the setting sun.

6.4 The chemical basis of vision. The curve in black shows the sensitivity of the (dark adapted) human eye to various wavelengths of light. The red dots show the amount of light over the same range of wavelengths absorbed by the photo-chemical rhodopsin in the frog's eye. The curves are substantially the same, indicating that the human eye, when dark adapted, functions by absorption of light by the same photochemical.

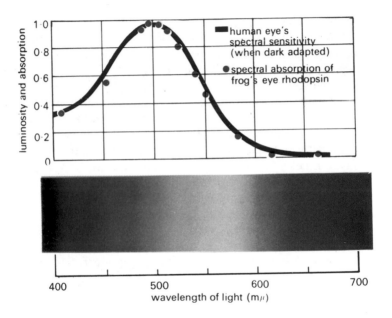

6.5 This shows how the sensitivity of the eye to various wavelengths of the spectrum is different when the eye is *light adapted*. The black curve shows the dark adapted sensitivity, while the red curve shows that this shifts along the spectrum with light adaptation, when the cones take over from the rods. This is known as the *Purkinje shift*.

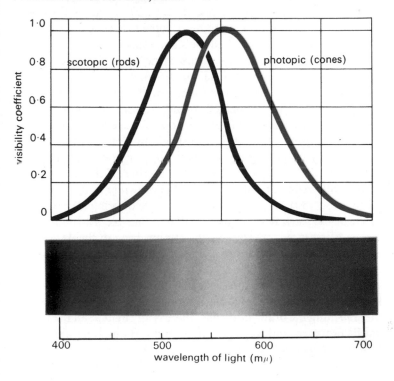

The eye's sensitivity to light

As intensity of light is increased the rate of firing of the receptors increases, intensity being signalled by the rate of firing. Unfortunately it is not possible to record the electrical activity in the receptors of a vertebrate eye because the retina is 'inside out' so that an electrode cannot reach them without doing extensive damage. By the time the optic nerve is reached the signals have been complicated by the cross-connections of the layers of nerve cells in the retina. There is however an eye – that of a living fossil, the Horse Shoe Crab, *Limulus*, found on the eastern seaboard of the United States – in which the receptors are connected directly to separate nerve fibres. The individual pathways of the ancient *Limulus* eye have turned out to be extremely useful; though, curiously enough, the creature seems to be effectively blind. Figure 6.6 shows the electrical activity in a nerve fibre of a *Limulus* eye. The rate of firing of its receptors is logarithmically related to intensity. This is shown in figure 6.8.

The first record (figure 6.6) shows a low rate of firing after the eye has been in the dark for one minute. The other record (figure 6.7) shows the firing rate increasing as the eye has been in the dark for a longer time. This corresponds to our own experience of increased brightness after being in the dark.

What happens when we look at a very faint light in an otherwise dark room? One might imagine that in the absence of light, there is no activity reaching the brain, and when there is any light at all the retina signals its presence, and we see the light. But it is not quite so simple. In the total absence of light, the retina and optic nerve are not entirely free of activity. There is some residual neural activity reaching the brain even when there is no stimulation of the eye by light. This is known from direct recording of the activity of the optic nerve in the fully dark-adapted cat's eye, and we have very strong reasons for believing that the same is true for the human and all other eyes.

This matter of a continuous background of random activity is of great importance, for it sets a continuous problem for the brain.

Imagine some neural pulses arriving at the brain: are they due to light entering the eye, or are they merely spontaneous 'noise' in the system? The brain's problem is to 'decide' whether neural activity is representing outside events, or whether it is mere 'noise', which

6.6 (Top) This shows the electrical activity, recorded on an oscilloscope, of a single fibre of the optic nerve of *Limulus* for three intensities of light. The rate of firing increases roughly in proportion to the log. of the intensity.
6.7 (Bottom) The rate of firing after various durations of darkness. With increasing dark adaptation the rate increases, corresponding to increase in apparent brightness though the actual intensity of the light is the same.

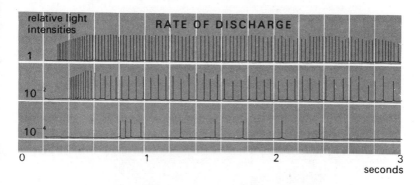

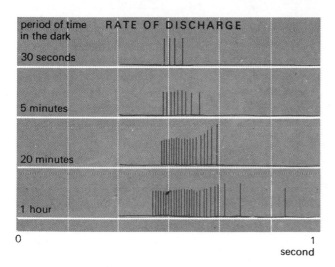

6.8 This is derived from the kind of records shown in 6.6 and 6.7. The rate of firing is plotted against the log. of the intensity, giving approximately a straight line, showing a logarithmic relation between rate of firing and intensity for constant adaptation.

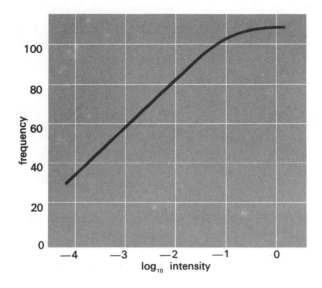

should be ignored. This is a situation very familiar to the communications engineer, for all sensitive detectors are subject to random-noise, which degenerates signals, and limits the sensitivity of detectors. There are ways of reducing harmful effects of noise: these are applied with good effect by radio astronomers, for noise masks the radio sources in space just as it masks and confuses weak visual signals. The eye adopts certain measures to reduce the effects of noise, notably increasing the duration over which signals are integrated (which we saw reflected in the Pulfrich Pendulum effect) and by demanding several confirming signals from neighbouring receptors serving as independent witnesses.

One of the oldest laws in experimental psychology is Weber's Law. This states that the smallest difference in intensity which can be detected is directly proportional to the background intensity. For

6.9 Weber's Law ($\Delta I/I = C$). Plotting ΔI against I, gives a horizontal straight line over a wide range of I, but the law breaks down at low intensities, when $\Delta I/I$ must be raised for detection. Plotting ΔI against I gives a substantially straight line down to small values of I, indicating a hidden constant k in the denominator. We may thus write the Law as $\Delta I/I + k = C$, where k appears to be related to neural noise level. These curves show the breakdown at low intensities.

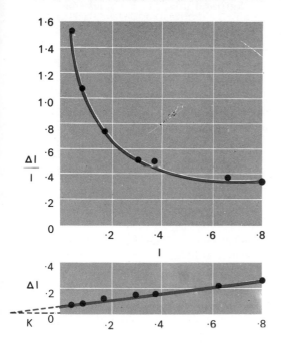

example: if we light a candle in a brightly lit room, its effect is scarcely noticeable; but if the room be dim to start with (lit by just a few other candles) then the added candle makes a marked difference. In fact we can detect a change in intensity of about one per cent of the background illumination. This is written $\dfrac{\Delta I}{I} = $ Constant (ΔI meaning the small incremental intensity over the background intensity, I). Now this law holds fairly well over a wide range of background intensity, I, but it breaks down for low intensities. This may be seen in figure 6.9 where, if Weber's Law did hold down to zero intensity, we would have a horizontal straight line, indicating invariance of the just detectable differential intensity $\frac{\Delta I}{I}$ over all values of I. In fact we get the full line shown in this graph, indicating a marked rise in $\frac{\Delta I}{I}$ as the background intensity becomes small. This breakdown is largely explained if we take into account the residual firing of the retinal cells in the absence of light. The residual activity is to the brain exactly equivalent to a more or less constant dim light added to the background. We may estimate its value by extrapolating back past the origin, and reading off the y-axis of the graph. This gives it in terms of an equivalent light intensity, we may call k. This hidden constant, k, can be attributed to the 'noise' of the retina. There is evidence that this internal noise of the visual system increases with age: The increased noise level is indeed no doubt partly responsible for the gradual loss of all visual discriminations with ageing. Increased neural noise may also affect motor control and memory.

The idea that discrimination is limited by noise in the nervous system has far-reaching consequences. It suggests that the old idea of a *threshold* intensity, above which stimuli need to go before they have any effect on the nervous system, is wrong. We now think of any stimulus as having an effect on the nervous system, but only being accepted as a signal of an outside event when the neural activity is unlikely to be merely a chance increase in the noise level. The situation may be represented as in figure 6.10. This shows a patch of light serving as a background (I) on which is added light (ΔI) being detected. These two intensities of light give rise to statistically distributed neural impulse rates. The problem for the brain is to 'decide' when a given increase is merely chance, and when it is due to the increased intensity of the signal. If the brain accepted *any* increase from the average activity, then we would 'see' flashes of light not in fact present half the time. We thus reach the idea that a statistically significant difference is demanded before neural activity is accepted as representing a signal. The smallest difference (ΔI) we can see is determined not simply by the sensitivity of the receptors of the retina, but also by the difference in neural pulse rate demanded for acceptance as a signal.

6.10 This attempts to show the statistical problem presented to the brain due to the random firing of nerves. When the signal field $(I + \Delta I)$ is being discriminated from its dimmer background (I) the pulse rates are not always different, but are distributed as in the graph. So we may see a 'light' due to 'noise', or miss it when the rate is lower than average. The brain demands a significant difference before accepting neural activity as a signal.

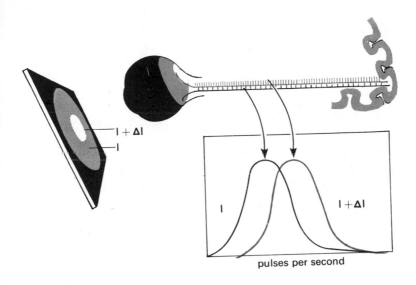

pulses per second

Sometimes we do see flashes which are not there. These are evidently due to the noise exceeding the demanded significance level, and this is bound to happen on occasions.

The choice of the level above which activity is accepted is a matter of trading reliability for sensitivity. There is evidence that this level is to some extent variable, and depends on our 'set'. When we are particularly careful, more information is demanded and sensitivity suffers.

This picture of intensity discrimination applies throughout the nervous system. It applies not only to differences between intensities but also to the absolute limit of detection against darkness, for the absolute threshold is determined by the smallest signal which can be detected reliably against the random background of the neural noise which is present in the visual brain though no light enters the eye.

Implications of randomness

Randomness – events occurring unpredictably without apparent cause – has been seized upon to rescue the nervous system (or rather our view of it) from being a machine without volition or Free Will. This takes us to most puzzling questions: – How can behaviour be understood or explained if there is a random element in the nervous system providing it? How can we be responsible for actions, either if the nervous system is precisely *determined* – or if it is partly *random*? It is sometimes argued that randomness allows Free Will: but how can we be responsible, or take credit, for actions set off by chance processes of the synapses? Here neurology meets philosophy and ethics: the debate continues.

There is a rather different, and less considered problem. It is generally assumed that there is a close tie-up between neural activity and consciousness. When we see a flash of light, we assume that somehow this is caused by cells of the nervous system sending signals which give this sensation. But if some of the activity is random – how can there be a simple one-to-one relation? This question probes deep into physiological psychology: for if there is no such simple relation we can hardly expect to relate neural activity in any direct way to experiences even as simple as flashes of light.

7 Seeing movement

Detection of movement is essential to survival. From the animals lowest on the evolutionary scale to man, moving objects are likely to be either dangerous or potential food, and so rapid appropriate action is demanded, while stationary objects can generally be ignored. Indeed it now seems that it is only the eyes of the highest animals which can signal to the brain in the absence of movement.

Something of the evolutionary development of the eye, from movement to shape perception, can be seen embalmed in the human retina. The edge of the retina is sensitive only to movement. This may be seen by getting someone to wave an object around at the side of the visual field, where only the edge of the retina is stimulated. It will be found that movement is seen but it is impossible to identify the object. When the movement stops, the object becomes invisible. This is as close as we can come to experiencing primitive perception. The very extreme edge of the retina is even more primitive: when stimulated by movement we experience nothing, but a reflex is initiated which rotates the eye to bring the moving object into central vision, so that the highly developed foveal region with its associated central neural network is brought into play for identifying the object. The edge of the retina is thus an early-warning device, used to rotate the eyes to aim the sophisticated object-recognition part of the system on to objects likely to be friend or foe or food rather than neutral.

Those eyes, like our own, which move in the head can give information of movement in two distinct ways. When the eye remains stationary, the image of a moving object will run across the receptors and give rise to velocity signals from the retinas; but when the eyes follow a moving object, the images remain more or less stationary upon the retinas, and so *they* cannot signal movement, *though we still see the movement of the object*. If the object is viewed against a fixed background, there will be velocity signals from the background, which now sweeps across the retinas as the eyes follow the moving object: but we still see movement even when there is *no visible background.* This can be demonstrated with a simple experiment. Ask someone to wave a lighted cigarette around slowly in a dark room; and follow it with the eyes. The movement of the cigarette is seen although there is no image moving across the retinas. Evidently, the rotation of the eyes in the head can give perception of movement, and fairly accurate estimates of velocity, in the absence of movement signals from the retinas.

There are then, two movement signalling systems. We will name them (a) The image/retina system, and (b) The eye/head system. (Figure 7.1.) (These names follow those used in gunnery, where similar considerations apply when guns are aimed from the moving platform of a ship. The gun turret may be stationary or following, but movement of the target can be detected in either case.)

We may now take a look at the image/retina system, and then see how the two systems work in collaboration.

The image retina movement system It is found, by recording the electrical activity from the retinas of animals, that there are various kinds of receptors, almost all signalling only changes of illumination and very few giving a continuous signal to a steady light. Some signal when a light is switched on, others when it is switched off, while others again signal when it is switched on or off. Those various kinds are named, appropriately enough, 'on', 'off', and 'on–off' receptors. It seems that those receptors responding only to change of illumination are responsible for signalling movement, and that *all eyes are primarily detectors of movement*. The receptors signalling only changes will respond to the leading and trailing edges of images, but will not signal the presence of stationary objects, unless the eyes are in movement.

By placing electrodes in the retinas of excised frogs' eyes, it has been found that analysis of the receptor activity takes place in the retina before the brain is reached. A paper charmingly titled: 'What the Frog's Eye Tells the Frog's Brain', by J. Y. Lettvin, H. R. Maturana, W. S. McCulloch and W. H. Pitts, of the Massachusetts Institute of Technology (MIT), describes retinal feature detectors which code aspects of the external world – which are the only features the brain receives from the frog's eyes. The brain is however not always involved: Horace Barlow at Cambridge had earlier dis-

7.1 a The image/retina system: the image of a moving object runs along the retina when the eyes are held still, giving information of movement through sequential firing of the receptors in its path.
b The eye/head movement system: when the eye follows a moving object the image remains stationary upon the retina, but we still see the movement. The two systems can sometimes disagree, giving curious illusions.

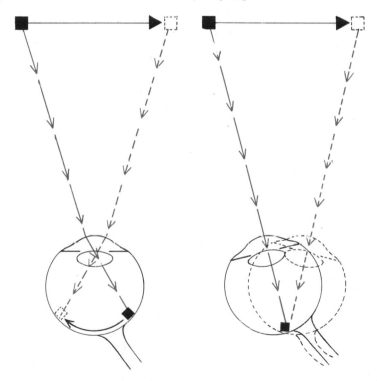

covered a retinal 'bug detector' in the frog's retina, eliciting reflex tongue fly-catching movement directly. The MIT group found:

1 Fibres responding only to sharply defined boundaries.

2 Fibres responding only to changes in the distribution of light.

3 Fibres responding only to a general dimming of illumination, such as might be caused by the shadow of a bird of prey.

The frog's eye signals only changing light patterns and moving contours. The frog's brain is fed only with these limited kinds of

7.2 Hermann von Helmholtz (1821–94), the greatest figure in the experimental study of vision. His *Physiological Optics* is still the most important work on the subject; indeed, disappointingly little has been added since.

visual information. Its brain is almost certainly too small and simple to infer much about objects from these meagre data.

It is now known that image/retina shifts are coded as movement at the retina of the rabbit (and presumably of man) while, as we have seen (Page 46), specific cells of the visual cortex of the cat and monkeys respond to movement (figure 4.6).

The physiological discovery that movement is coded into neural activity in the retina, or immediately behind the retina in the visual projection areas of the brain, is important in many respects; but in particular it shows that *velocity can be perceived without involving an estimate of time*. It is, however, sometimes assumed that the neural systems giving perception of velocity must resort to an internal 'biological clock'. Velocity is defined in physics as the time taken for an object to travel a given distance ($v = \dfrac{d}{t}$). It is then assumed that a time estimate is always required to estimate velocity. But the speedometer of a car has no clock associated with it. A clock is needed for calibrating such an instrument in the first place, but once calibrated it will give velocity measures without the use of a clock, and the same is evidently true of the eye. The image running across the retina sequentially fires the receptors, and the faster the image travels, up to a limit, the greater the velocity signal this gives. Analogies with other velocity detectors (speedometers and so on) show that velocity could be perceived without reference to a 'clock', but they do not tell us precisely how the visual movement system works. A model has been suggested for the compound eye of beetles. This has been made and is used in aircraft to detect drift due to wind blowing the machine off course. This movement detector was developed by biological evolution some hundred million years ago, discovered by applying electronic ideas, and then built for flight by man.

The eye/head movement system The neural system giving perception of movement by shift of images across the retina must be very different from the way movement is signalled by rotation of the eyes in the head. Somehow, the fact that the eye is being moved is signalled to the brain, and used to indicate the movement of external objects. That this does happen is demonstrated with the cigarette experiment we have just described, for in that situation there is no systematic movement across the retina, and yet the movement of the cigarette is seen when it is followed by the eyes (figure 7.1b).

The most obvious kind of signal would be from the eye muscles, so that when they stretch signals would be fed back to the brain indicating movement of the eyes, and so of objects followed by the eyes. This would be the engineer's normal solution; but is it Nature's? We find the answer when we look at what may seem to be a different question.

Why does the world remain stable when we move our eyes? The retinal images run across the receptors whenever we move our eyes, and yet we do not experience movement – the world does not usually spin round when we move our eyes. How can it remain stable?

We have seen that there are two neural systems for signalling movement, the *image/retina* and the *eye/head* movement systems. It seems that during normal eye movements their signals cancel each other out – to give stability to the visual world. The idea of cancellation to give visual stability was discussed by the physiologist who did most to unravel spinal reflexes, Sir Charles Sherrington (1857–1952), and by Helmholtz; but they had different ideas as to how it comes about, and especially how what we have called the '*eye/head*' movement system conveys its information. Sherrington's theory is known as the *inflow theory* and Helmholtz's as the *outflow theory* (figure 7.3). Sherrington thought that signals from the eye muscles are fed back into the brain when the eye moves, to cancel the movement signals from the retina. This principle is familiar to engineers as feed-back. But neural signals from the eye muscles would take a considerable time to arrive. So we might expect a sickening jolt, just after movements of the eyes; before the inflow signals reach the brain to cancel the *image/retina* movement signals. Helmholtz had a very different idea. He thought that the retinal movement signals are cancelled *not* by 'inflow' signals from the eye muscles – but by central 'outflow' signals from the brain commanding the eyes to move.

The issue can be decided by very simple experiments, which the reader can try himself. Try pushing an eye gently with the finger, while the other is closed by holding a hand over it. When the eye is

7.3 Why does the world remain stable when we move our eyes? The *inflow theory* suggests that the movement signals from the retina (image/retina system) are cancelled by (afferent) signals from the eye muscles. The *outflow theory* suggests that the retinal movement signals are cancelled by the (efferent) command signals to move the eyes, through an internal monitoring loop. The evidence favours the outflow theory.

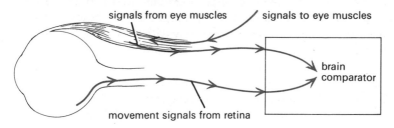

signals from eye muscles   signals to eye muscles

brain
comparator

movement signals from retina

INFLOW THEORY

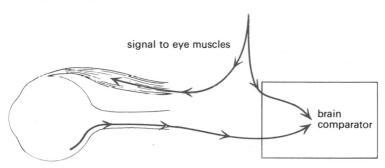

signal to eye muscles

brain
comparator

OUTFLOW THEORY

rotated passively in this way, the world will be seen to swing round in the opposite direction to the movements of the eyes. Evidently stability does not hold for *passive* eye movements, though it does for the normal *voluntary* eye movements. Since the world swings round *against* the direction of the passive eye movements, it is evident that the image/retina system still works; it is the eye/head system which is not operating. We may well ask: why should the eye/head system work only for voluntary and not for passive eye movements? Sherrington thought that it works by sending down signals from stretch receptors in the eye muscles. Such stretch receptors are known to give feed-back signals from the muscles which move the limbs. But it looks as though the eye/head system does not work this way, for the stretch receptors should surely deliver signals during the passive movements of the eyes.

We may stop all retinal movement signals, and see what happens during passive movements of the eyes. This is easily done by staring at a bright light (or a photographic flash) to get an after-image. This produces a local area of fatigue, like a photograph stuck on the retina, and it will move precisely with the eye – and so cannot give any image/retina movement signals however the eye moves. If we observe the after-image in darkness (to avoid a background), it is found that if the eyes are pushed with the finger to move passively *the after-image does not move.* This is very strong evidence against the inflow theory, for stretch receptor activity should cause the after-image to shift with the eyes; if it is normally responsible for cancelling the retinal movement signals.

When you move your eyes voluntarily, you will see that the *after-image moves with the eyes.* Wherever the eyes are moved, the after-image will follow. Helmholtz explained this by supposing that it is not activity *from* the eye muscles which is involved, but *commands* to move the eyes. This outflow theory suggests that the command signals to move the eyes are monitored by an internal loop (figure 7.3) and that these monitored signals cancel the image/retina signals. When these are absent, as in the case of after-images viewed in darkness, the world swings round with the eyes because the command signals are not cancelled out by retinal signals. Passive movements of the eye give no movement of after-images – for neither system gives a movement signal.

In clinical cases where something has gone wrong with the eye muscles or their nerve supply, the world swings round for these patients when they try to move their eyes. Their world moves in the direction their eyes should have moved. This also occurs if the muscles are prevented from functioning, by *curare*, the South American arrow poison. The Austrian physicist Ernst Mach (1838–1916) bunged up his eyes with putty, so they could not move, and he got the same result.

The eye/head system, then, does not work by actual movement of the eyes, but by commands to move them. It works even when the eyes do not obey the commands. It is surprising that command signals can give rise to perception of movement: we usually think of movement perception as always coming from the eyes, not from a source deep in the brain controlling them.

Why should such a peculiar system have evolved? It is even more surprising when we find that there are in fact stretch receptors in the eye muscles. An inflow or feed-back system would appear to be too slow: by the time a feed-back signal got back to the brain for cancelling against the retinal movement signal it would be too late.

The cancelling signal could start at the same moment as the command, and so could oppose the retinal signal with no delay. Actually, the signal from the retina takes a little time to arrive (the 'retinal action time'), and so the command signal could arrive for cancelling too soon; but it is delayed to suit the retina, as we may see by studying carefully the movement of the after-image with voluntary eye movements. Whenever the eye is moved, the after-image takes a little time to catch up, and this is evidently the delay put into the monitored command signal so that it does not arrive before the signal from the retina. Can one imagine a more beautiful system?

Illusions of movement

We may now look at some illusions of movement. Like other illusions they can be of practical importance, and they can throw light on normal processes.

The case of the wandering light   The reader might like to try the following experiment. The apparatus is a lighted cigarette placed on an ash-tray at the far end of a completely dark room. If the glowing end is observed for more than a few seconds, it will be found to wander around in a curious erratic manner, sometimes swooping in one direction, sometimes oscillating gently to and fro. Its movement may be paradoxical; it may appear to move and yet not to change its position. This perceptual paradox is important in understanding not only this phenomenon of the light that moves, but also the very basis of how movement is represented and coded in the nervous system.

This effect of the light that moves in the dark is known as the *autokinetic phenomenon*. It has received a great deal of discussion and experimental work. A dozen theories have been advanced to explain

101

it, and it has even been used as an index of suggestibility and group interaction: for people tend to see it moving in the same direction that other people present claim to see it moving – though of course it is in fact stationary.

The theories to explain the effect are extraordinarily diverse. It has been suggested that small particles floating in the aqueous humour, in the front chamber of the eye, may drift about and be dimly seen under these conditions. It is then supposed that the spot of light and not the particles *seems* to move, just as the moon may appear to scud through the sky on a night when the clouds are driven fast by the wind. This effect – called 'induced movement' – will be discussed later (pages 114–15). There is however plenty of evidence that this is not responsible for the autokinetic effect, for the movements occur in directions unrelated to the drift of the particles in the eye (when these are made more clearly visible with oblique lighting) and in any case they are not generally visible. A further theory (which is sometimes held by ophthalmologists) is that the eyes cannot maintain their fixation accurately on a spot of light viewed in darkness, and that the drifting of the eyes causes the image of the spot of light to wander over the retina, causing the apparent movements of the light. This theory was all but disproved in 1928 by Guilford and Dallenbach, who photographed the eyes while the subjects observed the spot of light, and reported what movements they saw. The reported movements of the spot were compared with the photographic records of the eye movements, and no relation was found between the two. In addition, the eye movements under these conditions were extremely small. This experiment seems to have gone largely unnoticed.

All attempts but one to explain the wandering of the light in the dark suppose that *something* is moving – the particles in the aqueous humour, the eyes, or some sort of internal reference frame. The last suggestion formed an important part of the Gestalt psychologists' theory of perception. They attached great weight to the wandering-light effect. Kurt Koffka, in his celebrated *Principles of Gestalt Psychology* of 1935, says of it:

These 'autokinetic movements', then, prove that no fixed retinal values belong to retinal points; they produce localisation within a framework, but do so no longer when the framework is lost. . . . The autokinetic movements are the most impressive demonstration of the existence and functional effectiveness of the general spatial framework, but the operation of this framework pervades our whole experience.

Is this argument sound? I believe it contains an important fallacy. What is true for the world of objects, and their observations, does not necessarily hold for *errors* of observation, or illusions. It is important to appreciate the difference. Any sense organ can give false information: pressure on the eye makes us see light in darkness; electrical stimulation of any sensory endings will produce the experience normally given by that sense. Similarly, if movement is represented in neural pathways, *we should expect to experience illusions of movement if these pathways are activated or disturbed.* This is familiar from man-made detectors of movement: the speedometer of a car may become stuck at a reading of, say, 50 km.p.h. and will indicate this speed though the car is travelling at 100 km.p.h. – or is stationary and locked up in its garage.

The confusion, and it is a serious and common confusion, has I believe arisen through a failure to distinguish between the conditions necessary for *valid* estimates of the velocity of objects, and those holding for *invalid* estimates – illusions.

It is true that all real movement of objects in the world is relative, and we can only speak of, or measure, the movement of objects by reference to other objects. This is indeed the basis of Einstein's Special Theory of Relativity. This point was however made two hundred years before Einstein, by Bishop George Berkeley, when he challenged a point in Newton's *Principia* of 1687:

If every place is relative, then every motion is relative. . . . Motion cannot be understood without a determination of its direction which in its turn cannot be understood except in relation to our or some other body. Up, Down, Right, Left, all directions and places are based on some relation and it is necessary to suppose another body distinct from the moving one . . . so that motion is relative in its nature . . . .

Therefore, if we suppose that everything is annihilated except one globe, it would be impossible to imagine any movement of that globe.

But it has been assumed by writers on perception that if nothing is moving – not the eyes, particles in the eyes, nor anything else – it would be impossible to experience even *illusions* of movement, for

103

7.4 These 'clock histograms' show how a small dim light viewed in darkness appears to move after straining the eyes in four different directions for 30 seconds each time. The arrows show the direction of strain; the dark tinted areas show the direction of apparent movement for the next 30 seconds, while the light tinted areas show the movement during the following 30 seconds. The numbers give the duration in seconds of movement over two minutes after strain.

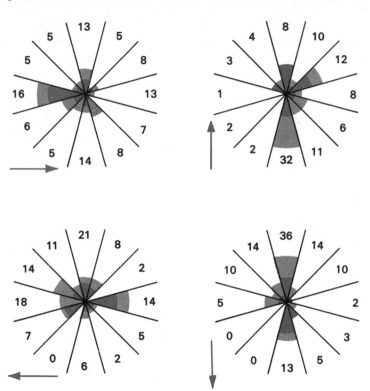

example, of the spot of light in darkness. The wandering light has been taken to represent the same situation as Berkeley's globe when everything except it is annihilated, but it is very different.

The error lies in supposing that *false* estimates of movement, or illusions of movement, require something moving relative to something else. But they can result simply from a disturbance, or a loss of calibration, of the measuring instrument – whether it be a speedometer or the eye. We may now seek for the kind of disturbance or loss of calibration of the visual system which is responsible for the wandering light. To do this we shall try to produce systematic illusory movements of the spot of light; by deliberately upsetting the *eye/head* system.

If the eyes are held hard over for several seconds, in any direction, and then returned to their normal central position while the small dim light is viewed in darkness as before, the light will be seen to speed across in the direction in which the eyes were held – or sometimes in the opposite – but seldom in any other direction. The illusory movement may continue for several minutes when the eye muscles are asymmetrically fatigued (figure 7.4). Now fatigue of the eye muscles requires abnormal command signals to hold the eyes' fixation on the light, but these are the same as the signals which normally move the eyes when they follow a moving object. We thus see movement when the muscles are fatigued, although neither the eyes nor the image on the retinas are moving. The wandering illusory movements of the autokinetic effect seem to be due to the command signals maintaining fixation in spite of slight spontaneous fluctuations in the efficiency of the muscles, which tend to make the eyes wander. It is not the eyes moving, but the correcting signals applied to *prevent* them moving which cause the spot of light to wander in the dark.

We may now ask: if the correction signals move the spot of light in the dark, why do they not also cause instability under normal conditions? Why is the world generally stable? There is no certain answer to this question. It may be that in the presence of large fields of view the signals are ignored because the brain assumes that large objects are stable, unless there is clear evidence to the contrary. This is borne out by the effect of 'induced movement', which we shall discuss (page 114), but first we should recognise that sometimes the normal world does swing round.

The case of the wandering world The world swings round when we are fatigued or suffering from the less pleasant effects of alcohol. This was described by the Irish wit Richard Brinsley Sheridan. After a party, friends led him to the front door of his house, in Berkeley Square, and left him. Looking back, they saw him still standing in the same position. 'Why don't you go in?' they shouted. 'I'm waiting until my door goes by again . . . then I'll jump through!' replied Sheridan. Just how this ties up with the wandering spot of light is not entirely clear. It may be that the eye movement command system is upset, or it may be that alcohol serves to reduce the significance of the external world, so that error signals which are normally disregarded are accepted. Just as we can become possessed by fantasies and irrational fears when tired or drunk, so might we become dominated by small errors in the nervous system which are generally rejected as insignificant. If this is so one might expect schizophrenics to suffer from instability of their visual world; but I know of no evidence for this.

The waterfall effect We have found that the illusory movements of the spot of light viewed in darkness are due to small disturbances of the eye/head system. We might now expect to find illusions of movement due to disturbance of the image/retina system, and indeed we do. These illusions are not limited to movement of the whole field: various parts of the field may appear to move in different directions, and at different rates, the effects being bizarre and sometimes logically paradoxical. The most marked image/retina disturbance is known as the 'waterfall effect'.

The 'waterfall effect' was known to Aristotle. It is a dramatic example of illusory movement, caused by adaptation of the image/retina system. It may be induced most easily by looking steadily, for about half a minute, at the central pivot of a rotating record player. If the turntable is then stopped suddenly it will seem, for several seconds, to be rotating backwards. The same effect is found after looking at moving water, for if the eyes are then directed to the bank, or any fixed object, it will seem to flow in the direction opposite to the flow of water. The most dramatic effect is obtained from a rotating spiral (figure 7.5). This is seen to expand while rotating, and seems to contract as an after-effect when the spiral is stopped (or *vice versa* if the direction of rotation is reversed). This

7.5 When this spiral is rotated, it appears to shrink or expand, depending on the direction of rotation. But when stopped, it continues to *appear* to shrink (or expand), in the *opposite direction*. This cannot be due to eye movement, since the apparent shrinkage or expansion occurs in all directions at once. The effect is paradoxical — there is movement, but no change in position or size.

illusory contraction or expansion when the spiral is stopped cannot be due to eye movements, for the eyes can move only in one direction at a time, while the effect is a radial contraction or expansion occurring in all directions from the centre at the same time. This fact alone shows that we must attribute the effect to the image/retina, rather than to the eye/head, movement system. It is also quite easy to show conclusively that it is *solely* due to upset of the image/retina system. This can be shown by following a moving belt of stripes (figure 7.6) with the eyes, returning the eyes rapidly to the beginning (with the lights switched off) and following the movement again, several times. In this way movement in one direction is experienced using the eye/head but not the image/retina system. When the belt is stopped, after prolonged viewing with the following eye movements, there is *no* after-effect. So we attribute the waterfall effect to the image/retina system.

It remains a problem as to whether the adaptation takes place in the retina or in the brain. The retina seems rather too simple to be capable of such a complex after-effect, but it is very difficult to rule out retinal adaptation as a part-cause. One might think (and several psychologists who ought to have known better have thought) that the issue could be decided by looking at the moving stimulus object with one eye, while holding the other closed, and then observing whether the after-effect occurs when viewing a stationary object with the unstimulated eye. The answer is that it does occur, at about half strength. This does not, however, show conclusively that the adaptation took place in the brain, for it is possible that the stimulated eye goes on sending up a movement signal after it is shut and that this is, so to say, 'projected' into the field of the unstimulated eye. This is perfectly possible, for it is difficult to say which eye is active: one tends to think that whichever eye is open is doing the seeing. There have been experiments to try to decide the issue.

We have already seen from the work of Hubel and Wiesel that movement is represented in separate neural channels, and that different channels indicate different directions of movement (figure 4.6). It is reasonable to assume that these channels can become adapted, or fatigued, with prolonged stimulation (as happens with almost all other neural channels) and that this unbalances the system, giving illusory movement in the opposite direction.

7.6 The waterfall effect. This is similar to apparent movement induced with the rotating spiral. After watching the moving stripes they appear to have a backwards-going velocity when stopped. This only occurs when the movement is observed with the eye held stationary, and *not following* the stripes. It must be due to adaptation of the image/retina system only.

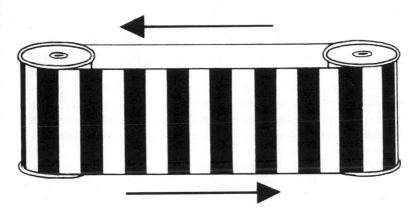

If the after-effect from the rotating spiral is examined carefully, two curious features will be noticed. The illusory movement may be paradoxical: it may *expand* and yet *not get any bigger*. Or, *shrink* but *not get any smaller*. It is changing and not changing. This sounds impossible, and it *is* impossible for real objects; but we must always remember that what holds for real objects may not hold for perception once we suffer illusions. We can experience things which are logically impossible when we suffer illusions. In this case of the after-effect of movement of an expansion with no change in size, we may suppose that this comes about because *velocity* and *positions* are indicated by separate neural mechanisms, and only one of these – the velocity system – is upset by continued viewing of the rotating spiral. This is like a trial judge getting incompatible evidence from two witnesses and accepting both stories, at least for a time, before deciding that one is probably correct and dismissing the other as a pack of lies which can best be ignored. The perceptual system of the eye and brain has many channels, and many sources of information: the brain must serve as judge. Sometimes different sources of incompatible information are at least for a time accepted together, and then we experience a paradox – things that cannot possibly occur together. We should not be too surprised that illusions and hallucinations of drugs are sometimes impossible to describe.

Linked adaptations A new phenomenon was discovered by Celeste McCollough, in 1965. She presented vertical black stripes on an orange ground, alternated with an identical grating of horizontal stripes on a blue ground. They were presented alternately, with slide projectors: five seconds for one set of stripes followed by one second of blackness, then five seconds of the other set of stripes of different colour and orientation. When the observers had been subjected to this alternation of stimuli, for from two to four minutes, black-and-white gratings of the same orientations were substituted for the coloured gratings. Celeste McCollough found that the black and white stripes now appeared coloured. The vertical stripes appeared with a blue-green ground, and the horizontal stripes appeared on an orange ground. These complementary colours were less saturated than the usual after-images of coloured lights, but were quite distinct. She suggested that this effect was due to colour adaptation of orientation-specific edge detectors. It has since been found that a very wide variety of visual stimuli (including movement) can be associated in this way with colour, or with many other visual dimensions. These after-effects are however always *negative*, as for the complementary colours which Celeste McCollough found in her original experiment.

These *contingent after-effects*, as they came to be called, show many of the characteristics of classical Pavlovian Conditioning. They build up gradually with repeated stimuli; they decay rapidly when elicited by the 'unconditioned stimuli' (the black and white stripes) though the effects persist for many hours, or even days, when left alone. It has been suggested that the links between the related stimuli are given by 'double duty' cells, teased out by this procedure. But it is now clear that highly 'artificial' – extremely unlikely – pairings of stimuli can occur; so this is hardly plausible as an explanation. We should not expect many cells to be pre-wired for highly unlikely combinations of stimuli, or events – such as movement in one specific direction with a particular colour. An alternative explanation is that this is a kind of perceptual learning. Why, though, should it always be 'negative'? It seems to be adaptation to produce *constancy* of perception, against

110

irrelevant sensory signals. We do not see colours (or indeed anything) as signalled by the receptors. We see colours corrected against the colour of the ambient light; and brightness is maintained constant against changes of intensity of the illumination, and so on. The same kind of corrections are applied to shape, movement, and to all sensory modalities. So there are no simple bricks for building perceptions. This makes perception peculiarly difficult to analyse, to describe, or relate closely to known physiological mechanisms. Celeste McCollough succeeded in fooling this compensating interactive system, which though difficult to analyse and describe adequately, generally makes the world less confused than it would appear as ever-changing patterns of unrelated patterns or stimuli. Subtle compensations are vital for perception. But when inappropriate, they generate illusions. Contingent after-effects are a new class of compensations which when inappropriate do not succeed – as normally perception does succeed – in leaping from stimuli to object perception but into error: which, however, we can use to gain understanding of how we see. We can learn a lot about perception from perceptual errors.

Movement in cinema and television As we have seen, all the sensory systems can be fooled. Perhaps the most persistent fooling is by the cinema and television. They present a series of still pictures – but what we see is continuous action. This relies on two distinct (though often confused) visual facts: *persistence of vision* and the *phi phenomenon*. Persistence of vision is simply the inability of the retina to follow and signal rapidly changing intensities. A light flashing at a rate greater than about 50 flashes per second appears steady: though for bright lights, and for the peripheral retina, this Critical Fusion Frequency (as it is called) may reach 100 or more flashes/second. Cinema presents 24 frames per second; but a three bladed shutter raises the flicker rate to 72 flashes/second, three for each picture. Television (British standard) presents 25 fresh pictures/second, each given twice, to raise the flicker rate to 50/second. Television flicker is reduced by 'interlaced scanning', in which horizontal sections of the picture are built up by scanning in bands of lines, rather than continuously down the screen. Nevertheless, television flicker can be annoying, and dangerous to sufferers from epilepsy. Indeed, flicker is used to evoke symptoms for diagnostic purposes. Flicker can also be a hazard when driving along a row of trees with a low sun, or when landing a helicopter in tropical conditions. The rotor blades can produce a violent flickering light which is disturbing and can be dangerous. The 'stroboscopes' used in discotheques can be similarly disturbing and should probably be avoided: as also, of course, should the dangerously high levels of sound which are quite capable of producing permanent hearing loss. The senses are surely too precious to be abused and damaged unnecessarily.

Low-frequency flicker produces very odd effects on normal observers as well as on those with a tendency to epilepsy. At flash rates of five-ten per second, brilliant colours and moving and stationary shapes may be seen and can be extremely vivid. Their origin is obscure, but they probably arise from direct disturbance of the visual system of the brain, the massive repeated bursts of retinal activity overloading the system. The patterns which are seen are so varied that it is difficult to deduce from their appearance anything about the kind of brain systems which have been disturbed. Stimulation by bright flashing lights can be an unpleasant experience leading to headache and nausea.

The other basic visual fact upon which the cinema depends is Apparent Movement, known as the 'phi phenomenon'. There is a vast literature of experimental studies on this effect. It is generally studied in the laboratory by using a very simple display – merely two lights which can be automatically switched so that just after one light has gone off the other comes on. What is seen – provided the distance between the lights and the time intervals between their flashes is about right – is a single light moving across from the position of the first light to the second. It was argued by the Gestalt psychologists that this apparent movement across the gap between the lights is due to an electrical charge in the brain sweeping across the visual projection area and filling in the gap. As it was thought that the phi phenomenon demonstrated such a process, it was studied intensively. Most authorities would now consider the Gestalt view of the matter mistaken. Consider once again the case of an image moving across the retina, giving perception of movement as a result of the sequential stimulation of the receptors between the two flashing lights, or between the separate cinema pictures of a moving object, when we still see movement – do we have to suppose some filling-in process? Is it not simply that the intermittent stimulus is adequate to actuate the retinal movement system, provided the gaps in space and time are not too great? The situation is like that of a key and a lock. A key does not have to be *exactly* a certain shape to work the lock. There is always some degree of tolerance. Indeed, some tolerance is essential, for otherwise any slight change in the lock or the key would prevent it working. It is most likely that the image-retina system operates with stimuli reasonably like those provided by the normal movements of retinal images from moving objects, but that it will tolerate intermittent images provided the jumps in space or time are not too large. The phi phenomenon does tell us something about the image/retina movement system: that it tolerates gaps, to maintain continuity as objects are hidden briefly behind obstructions, or retinal images behind blood vessels. As a fortunate pay off, this tolerence to gaps allows cinema and television to be economically possible.

The relativity of movement

So far we have considered the basic mechanisms for perceiving movement – either by stimulation of the retina by moving images, or by the eye following moving objects. There is, however, far more to the perception of movement. Whenever there is movement the brain has to decide what is moving and what is stationary, with respect to some reference frame. Although, as we have seen, it is fallacious to suppose that illusory movement necessarily involves any actual movement, it remains true that all real movement is relative, and a decision is always required. An obvious example occurs whenever we change our position – by walking or driving or flying. We generally know that the movement is due to our own movements among the surrounding objects, and not due to their movement, but this involves a decision. As we should expect, sometimes the decision is made wrongly, and then we get errors and illusions which can be particularly serious, because perception of movement is of prime biological importance for survival. This is as true in the case of man living in an advanced civilisation as ever it was in the primitive state.

Most perceptual research has been undertaken with the observer stationary – often looking into a box containing apparatus giving him flashing lights, or pictures of various kinds. But real-life perception occurs during free movement of the observer, in a world where some of the surrounding objects are also in motion. There are severe technical problems in investigating the real-life situation, but the attempt is well worth while. The results can be important for flying, driving and also for space flight. Astronauts had to be trained on highly realistic simulators before they could make accurate judgments of speed, size, size and distance in the alien lighting of the moon. Much the same is true also for pilots. For this reason, visual simulators are most important for flying and space training, and might be useful for driving. But simulators never copy reality perfectly: so skills developed by them are seldom quite the skills required for the real thing. Discrepancies may show up at awkward moments as 'Negative Transfer'.

As we have seen, there is always a decision involved as to just what is moving. If the observer is walking or running, there is generally not much of a problem, for he has a lot of information from his limbs, informing him of his movement in relation to the ground. But when he is carried along in a car or an aircraft the situation is very different. When he has his feet off the ground his only source of information is through the eyes; except during acceleration or deceleration when the balance organs of the middle ear give some, though often misleading, information.

The phenomenon known as *induced movement* was very thoroughly investigated by a Gestalt psychologist, K. Duncker. He devised several elegant demonstrations which show that when there is movement signalled only by vision, we tend to accept that it is the largest objects which are stationary, the smaller objects moving. A striking demonstration is given by projecting a spot of light on a large movable screen (figure 7.7). When the screen is moved, what is seen as moving is the spot of light, though in fact this is stationary. It should be noted that there actually is information available to the eye, for it is the image of the screen and not the spot which moves on the retina, but this information is not always sufficient to decide the issue. This is relevant to why it is that the visual world does not swing round with eye movements, (Cf. page 98).

It seems clear that since it is generally rather small objects which

7.7 Induced movement. A spot of light is projected on to a screen which is moved. It is the stationary spot which is seen as moving. This occurs when the moving part is larger, or more likely to be stationary. (After Duncker.)

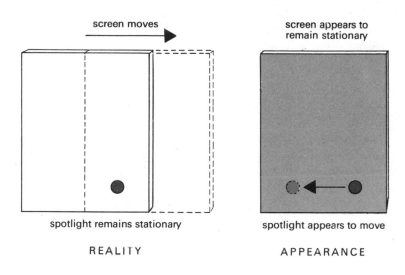

screen moves →

screen appears to remain stationary

spotlight remains stationary

spotlight appears to move

REALITY

APPEARANCE

move, the brain takes the best bet and tends to accept that movement is of smaller rather than larger objects when the issue can be in doubt. (The effect can be disturbing when driving a car – is it my hand-brake that is off, or is it that idiot over there running backwards? The answer is very often important!)

Apparent movement and distance When we observe the moon or stars while travelling in a car, we see them apparently moving along with us rather slowly. At fifty km. per hour, the moon may seem to move at ten–twenty km. per hour. We see it as moving more slowly than us, but keeping up, never falling behind. This is a curious effect.

The moon is so distant that we can regard it as being at infinity. As the car moves along, the angle to the moon remains virtually unchanged – it does not change its position although we are moving along under it. But perceptually it lies at a distance of only a few hundred metres. We know this from its apparent size. It subtends an

115

angle of $\frac{1}{2}°$, but looks the size of an orange, or sometimes a football, a few hundred metres away. Now consider this comparison object, a few hundred metres away, looking the same size as the moon. If we drive past *it* we rapidly leave it behind. But the moon does not get left behind, because it is in fact so distant, and the only way the perceptual system can reconcile these facts is to interpret them as an object moving with the car. The apparent velocity of the moon is determined by its apparent distance. If we change the apparent distance of the moon, by viewing it through converging prisms to change convergence of the eyes, then it seems to move at a different speed.

A related effect is observed with stereoscopic projection of lantern slides. If a scene is projected in stereo depth, using cross polaroids, it seems to rotate – to follow the observer as he moves. Thus a picture in 3-D of a corridor swings round so that the apparent front moves with the observer, the corridor seemingly aimed at him. This effect is disturbing and can produce nausea. If convergence of the eyes is set parallel, to infinity, the entire scene from front to back shifts *with the observer* as he moves. The effect is opposite to normal motion parallax; for the world rotates around the point of fixation *against*, not *with* the observer as he moves.

Stereo projection is particularly interesting in this connection because the objects observed in fact lie flat on the screen, although they are seen in depth, and so we have the situation of observer-movement with no physical motion parallax. Normally, when we move sideways, say to the right, nearer objects move to the left. Geometrically the world swings round the point of fixation of the eyes, against our movement. But when we observe pictures in stereo depth, the exact opposite happens; they appear to rotate *with* the observer's movement, the point of rotation being determined by the convergence of the eyes. This is set not at the will of the observer, but by the separation of the stereo pairs on the screen. (Given a stereo projector, these effects are well worth seeing.)

When the observer is carried along with his feet off the ground, he has to rely on vision to know that he is moving, and to judge his speed. In an aircraft there is little or no sensation of movement when flying high, and at landing and take-off it is a toss-up whether we see ourselves moving, or the ground rushing up to meet us. Illusions and errors in this situation are frequent and dramatic. So much so that pilots have to learn to dispense very largely with their normal perception, and to rely on instruments.

The situation is similar to that of induced movement. We make the best bet on very little evidence. The main evidence under normal conditions is systematic movement right across the retina, particularly at the periphery. If, for example, a rotating spiral, like figure 7.5, is filmed and projected very large on a cinema screen, we seem to be moving forwards, or away, from it – rather than seeing it expanding or contracting as when it fills but part of the eye. The entire retina seldom receives systematic movement except when the observer is moving, carrying his eyes through object-filled space. This is why flight simulators need large displays to be effective; and wide screen cinema can sweep us from our seats with moving images.

8.1 Thomas Young (1773–1829), after Lawrence. He was the founder with Helmholtz, of modern studies of colour vision. A universal genius, Young made important contributions to science and also to Egyptology, helping to translate the Rosetta Stone.

8 Seeing colour

The study of colour vision is an off-shoot from the main study of visual perception. It is almost certain that no mammals up to the Primates possess colour vision – if some do it is extremely rudimentary. What makes this so strange is that many lower animals do possess excellent colour vision: it is highly developed in birds, fish, reptiles and insects such as bees and dragon flies. We attach such importance to our perception of colour – it is central to visual aesthetics and profoundly affects our emotional state – that it is difficult to imagine the grey world of other mammals, including our pet cats and dogs.

The history of the investigation of colour vision is remarkable for its acrimony. The problems have aroused lifelong passions: the story being told of a meeting of fifty experts on colour vision who defended fifty-one theories! Perhaps some theories never quite die; but it now seems that the first suggested is essentially correct.

The scientific study of colour vision starts with Newton's great work, the *Opticks* (1704). This is surely the scientific book of its period most worth reading today. The extraordinarily imaginative and painstaking experiments were carried out in Newton's rooms in Trinity College, Cambridge, which still exist and are still lived in. It was there that his pet dog Diamond upset a lighted taper while Newton was in chapel, burning his chemical laboratory (in which he tried to turn common metals into gold) and some prized manuscript notes on his optical experiments. He delayed publishing *Opticks* until after the death of a rival genius, Robert Hooke, whose *Micrographia* (1665) is also well worth reading today. The *Opticks* contains the 'Queries', which are Newton's final and most speculative thoughts on physics and man's relation through perception with the universe.

Newton showed that white light is made up of all spectral colours, and with the later development of the wave theory of light it became clear that each colour corresponds to a given frequency. The essential problem for the eye, then, is how to get a different neural response for

119

different frequencies. The problem is acute because the frequencies of radiation in the visible spectrum are so high – far higher than the nerves can follow directly. In fact the highest number of impulses a nerve can transmit is slightly under 1,000/second, while the frequency of light is a million million cycles/second. The problem is: How is frequency of light represented by the slow-acting nervous system?

This problem was tackled for the first time by Thomas Young (1773–1829) who suggested the theory, further developed by Helmholtz, which is still the best we have. Young's contribution was assessed by Clerk Maxwell, in the following words:

It seems almost a truism to say that colour is a sensation; and yet Young, by honestly recognising this elementary truth, established the first consistent theory of colour. So far as I know, Thomas Young was the first who, starting from the well-known fact that there are three primary colours, sought for the explanation of this fact, not in the nature of light but in the constitution of man.

If there were receptors sensitive to every separable colour, there would have to be at least 200 kinds of receptor. But this is impossible – for the very good reason that we can see almost as well in coloured as in white light. The effective density of the receptors cannot therefore be reduced very much in monochromatic light, and so there cannot be more than a very few kinds of colour-responsive receptors. Young saw this clearly. In 1801 he wrote:

Now, as it is almost impossible to conceive each sensitive point of the retina to contain an infinite number of particles, each capable of vibrating in perfect unison with every possible undulation, it becomes necessary to suppose the number limited, for instance, to the principal colours, red, yellow and blue . . .

Writing a little later, he stuck to the number of 'principal colours' as three; but changed them from red, yellow and blue, to red, green and violet.

We have now come to the hub of the problem: How can all the colours be represented by only a few kinds of receptor? Was Young right in supposing the number to be three? Can we discover the 'principal colours'?

The possibility that the full gamut may be given by only a few 'principal' colours is shown by a single and basic observation:

colours can be mixed. This may seem obvious, but in fact the eye behaves very differently in this respect from the ear. Two sounds cannot be mixed to give a different pure third sound, but two colours give a third colour in which the constituents cannot be identified. Constituent sounds are heard as a chord, and can be separately identified, at any rate by musicians, but no training allows us to do the same for light.

We should be very clear at this point just what we mean by mixing colours. The painter mixes yellow and blue to produce green, but he is not mixing lights; he is mixing the total spectrum of colours *minus the colours absorbed by his pigments*. This is so confusing that we will forget about pigments, and consider only mixing of coloured *lights*, which may be produced by filters, or by prisms or interference gratings (see figure 8.2).

Yellow is obtained by combining red with green light. Young suggested that yellow is always seen by effective red–green mixture, there being no separate type of receptor sensitive to yellow light, but rather two sets of receptors sensitive respectively to red and green, the combined activity of which gives the sensation yellow.

Perhaps the fulcrum of controversies over colour theories is the perception of yellow. Is yellow seen by combined activity of red + green systems, or is it *primary* – as its simple perceptual quality might suggest? Although the *pure* appearance of yellow – for indeed it does not look like a mixture – has been raised against Thomas Young, this argument is not valid. The fact is that when a red and a green light are combined (by projecting these lights on a screen) we do see yellow, and the sensation is indistinguishable from that given by monochromatic light from the yellow region of the spectrum. It is certain that in this instance simplicity of sensation is no guide to simplicity of the underlying neural basis of the sensation; and it seems that this is generally true for all sensations and perceptions.

Young chose three 'principal colours' for a very good reason. He found that he could produce any colour visible in the spectrum (and white) by mixture of three, but not less than three, lights set to appropriate intensities. He also found that the choice of suitable wavelengths is quite wide, and this is why it is so difficult to answer the question: what are the principal colours? If it were the case that only three particular colours would give by mixture the range of spectral

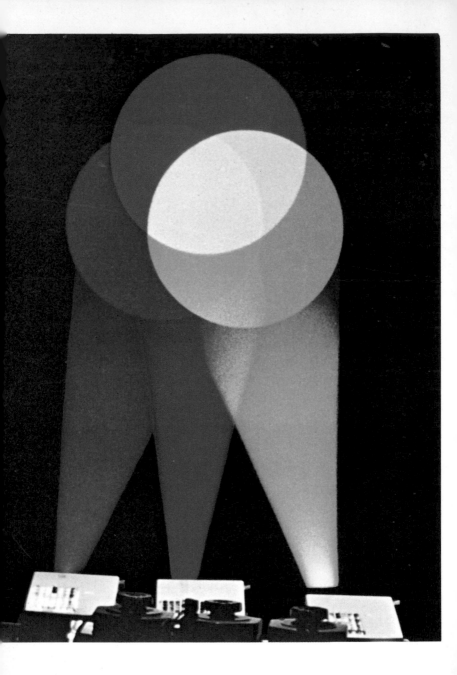

8.2 (Left) Young's experiment on colour mixture. By mixing three lights (not pigments) widely spaced along the spectrum, Young showed that any spectral hue could be produced by adjusting the relative intensities suitably. He could also make white, but not black or non-spectral colours such as brown. He argued that the eye effectively mixes three colours, to which it is basically sensitive. This remains the key idea in explaining colour vision.

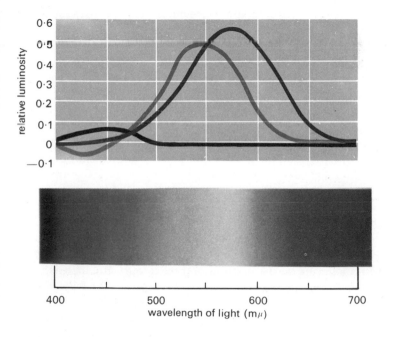

8.3 (Above) The fundamental colour response curves of the eye, according to W. D. Wright. These represent the supposed absorption curves of the three colour-sensitive pigments. all colours being seen by effective mixture of these.

hues, then we could say with some certainty that these would correspond to the basic colour systems of the eye, but there is no unique set of three wavelengths which will do the trick.

Young's demonstration is very beautiful. Figure 8.2 gives an idea of what it looks like.

The Young–Helmholtz theory is, then, that there are three colour-sensitive kinds of receptor (cones) which respond respectively to red, green and blue (or violet), and that all colours are seen by mixture of signals from the three systems. A great deal of work has gone into trying to isolate the basic response curves, and this has proved surprisingly difficult. The best established curves are shown in figure 8.3.

We may now look at a further graph, and one fundamental to the understanding of colour vision – the so-called hue-discrimination curve (figure 8.4). This compares wavelength of light with the smallest difference which produces a difference in hue. Now if we look at the earlier graph (figure 8.3) we see that hue should change very little as wavelength is varied from the ends of the spectrum, for the only change there is a gradual increase in the activity of the red or the blue systems, with no other system coming into play. We should see at the ends of the spectrum a change in brightness but not in colour. This is what happens. In the middle of the spectrum, on the other hand, we should expect dramatic changes in colour as the red system rapidly falls in sensitivity and the green rapidly rises – a small shift of wavelength should produce a large change in the relative activities of the red and green systems, giving a marked change of hue. We should thus expect hue discrimination to be exceptionally good around yellow – and this is indeed the case.

We shall pass over the later acrimonious debates on whether there are three, four, or seven colour systems, and accept Young's notion that all colours are due to mixtures of three colours. But there is more to colour vision than that revealed by experiments with simple coloured patches. Recently a jolt has been given to the more complacent by the American inventive genius Edwin Land. Apart from inventing Polaroid, when a research student, and later developing the Land camera, he has shown with elegant demonstrations that what is true for colour mixture of simple patches of light is not the whole story of the perception of colour. Odd things happen when the patches are more complicated, and represent objects. What Land has recently shown has been known in general for many years, but to him belongs the credit of emphasizing the additions to colour experience brought about by the more complicated situations of photographs and real objects. Indeed his work serves to remind us of the dangers of losing phenomena through simplifying situations in order to get neat experiments.

What Land did was to repeat Young's colour mixture experiment, but using not simple patches of light but photographic transparencies. Now, we may think of all projection colour photography as essentially Young's experiment put to work, for colour films provide physically only three colours. Land simplified it to two, and found that a surprising wealth of colour is given by only two light wavelengths when these form patterns or pictures. The technique is to take two photographic negatives of the same scene, each through a different colour filter. The negatives are converted into positive transparencies, and projected through their original filters, to give superimposed pictures on the screen. Quite good results are obtained simply with a red filter for one projector and no filter for the other. Now on Young's experiment, we would expect nothing but pink, of varying saturation (amount of added white) but instead we get green and other colours not physically present. This kind of result might, however, have been anticipated from two well known facts. First, the early colour films used only two colours, but it was not realised how good they could be. Secondly, as we have seen, although Young found that the *spectral* hues, and white, could be produced by mixture of three lights, it is *not* possible to produce *any* colour that can be seen. For example, brown cannot be produced, and neither can the metallic colours, such as silver and gold. So there is something odd about three colours, let alone two.

Consider an ordinary kodachrome colour transparency projected on to a screen. This gives us all the colours we ever see, and yet it consists of only the three lights of Young's experiment. The colour film is no more than a complex spatial arrangement of three coloured filters, and yet this gives us brown and the other colours Young was unable to produce with his three colours. It seems that when the three lights are arranged in complex patterns, and especially when they represent objects, we see a greater wealth of colour than when the same lights are present as a simple pattern, as in figure 8.2.

8.4 Hue discrimination curve. This shows how the smallest difference in wavelength ($\Delta\lambda$) varies with the wavelength of light (λ). It should be smallest (best colour discrimination) where the fundamental response curves (figure 8.3) have their steepest slopes. Roughly, this is true.

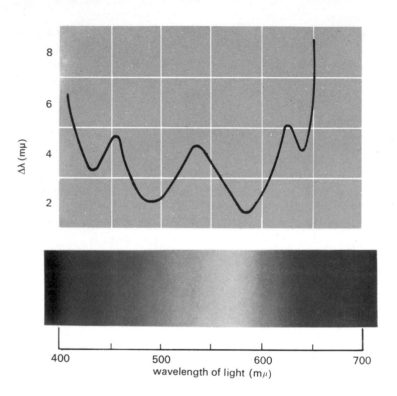

This means that any simple account of colour vision is doomed to failure: colour depends not only on the stimulus wavelengths and intensities, but also on differences of intensity between regions, and whether the patterns are accepted as representing objects. This involves high-level processes in the brain, which are extremely difficult to investigate. Brown normally requires contrast, pattern, and preferably interpretation of areas of light as surfaces of objects (such as wood) before it is seen, and yet in normal life brown is one of the most common colours.

The eye tends to accept as white not a particular mixture of colours, but rather the general illumination whatever this may be. Thus we see a car's headlamps as white while on a country drive, but in town where there are bright white lights for comparison, they look quite yellow; and the same is true of candle or lamplight. This means that the reference for what is taken as white can shift. Expectation or knowledge of the normal colour of objects is important. It is probable that objects such as oranges and lemons take on a richer and more natural colour when they are recognised as such, but this is certainly not the whole story. Land was careful to use objects whose colours could not have been known to the observers – objects such as reels of plastic-covered wire, and materials having woven patterns in coloured wool. He still got dramatic results.

Whatever the final assessment, and there is wide divergence of opinion, it is clear that Land's work brings out the complicated additions made by comparisons across the retina, and by the brain, when sensations are organised into perceptions of objects. It is all too easy in thinking about vision to concentrate on the eye and forget the brain and its stored knowledge and predictions from limited data.

Colour blindness

It is quite remarkable that even the common form of colour confusion – red confused with green – was not discovered before the late eighteenth century, when the chemist John Dalton found that he could not distinguish certain substances by their colours although other people could do so without difficulty. The reason is no doubt in part that we name objects by a variety of criteria. We call grass green, though we have no idea whether the sensation is the same for different people. Grass is a certain kind of plant found on lawns, and the sensation of colour which it gives we all call 'green', but we identify grass by other characteristics than its colour – the form of the leaves, their density and so on – and if we do tend to confuse the colour there is generally sufficient additional evidence to identify it as grass. We know it is supposed to be green, and we call it green even when this may be doubtful, as in dim light.

In the case of a chemist identifying substances, however, there are occasions when it is *only* the colour of the substance, in its bottle, which he can use for identification, and then his ability to identify and name colours as such will be put to the test. Tests of colour vision all depend on isolating colour as the one identifying characteristic, and then it is easy to show whether an individual has normal ability to distinguish between colours, or whether he sees as a single colour what to other people appear different colours.

The most common colour confusion is between red and green, as we have said, but there are many other kinds of confusion. Red–green confusion is surprisingly common. Nearly ten per cent of men are markedly deficient, though it is extremely rare in women. Less common is green–blue confusion. Colour blindness is classified into three main types, based on the supposed three receptor systems. They used to be called simply Red-, Green- and Blue-Blindness, but the colour names are now avoided. Some people are completely lacking in one of the three kinds of cone systems – they are now called *protanopes*, *deuteranopes* and *tritanopes* (after the first, second and third colour-sensitive systems), but this does not clarify the situation very much. These people require only two coloured lights to match all the spectral colours *they* can see. Thus Young's colour mixture result applies only to most people – not to extreme cases of colour blindness. It is more common to find not a complete absence of a colour system, but rather a reduced sensitivity to some colours. These are classed as *protanopia*, *deuteranopia* and *tritanopia*. The last, tritanopia, is extremely rare. People with these deficiencies are described as having *anomalous colour vision*. This means that although they require mixtures of three coloured lights to make their spectral colours, they use different proportions from the normal.

The proportions of red and green light required to match a mono-chromatic yellow is the most important measure of colour anomaly.

It was discovered (by Lord Rayleigh, in 1881) that people who confuse red with green, require a greater intensity either of red, or of green, to match yellow. Special instruments are made for testing colour vision, which provide a monochromatic yellow field placed close to a red-plus-green mixture field. The relative intensities of the red and green in the mixture can be varied, until the mixture gives the same colour to the observer as the monochromatic yellow. The proportions are read off a scale, which indicates the degree of protanopia or deuteranopia. The instrument is called an anomaloscope.

Yellow seems such a pure colour that it has often been thought that there must be a special 'yellow' set of receptors. But it can be shown quite simply, with the anomaloscope, that yellow is in fact always seen by effective mixture of 'red' and 'green' receptors.

An observer adjusts an anomaloscope so that he sees an identical yellow in the mixture and the monochromatic fields. He then looks into a bright red light, to adapt the eye to red. While the retina is adapted to red, he looks back into the anomaloscope, and is asked to judge whether the two fields still look the same colour. *He will see both fields as green*, and they will be the *same* green. The match is *not* upset by the adaptation to red, and so he will not require a different proportion of red and green light in the mixture field to match the monochromatic yellow. It would therefore be impossible to tell from the setting of the anomaloscope that he has been adapted to red: though what he sees is quite different when adapted – a vivid green instead of yellow. The same is true for adaptation to green light: both fields will then look the *same* red. The match still holds with adaptation to red or green light (figure 8.5).

But if there were a separate yellow receptor, this could not happen. A separate yellow receptor would allow the monochromatic field to be seen as yellow inspite of adaptation to red or to green, which must change the *mixture* field. A simple receptor could not get pushed along the spectrum scale by adaptation. But yellow seen by the mixture of red and green receptor systems should be shifted, if the sensitivity of the red or the green systems is affected by the adapting light, for this is equivalent to changing their relative intensities in the mixture field. There cannot be a different system operating for the two fields, given that each is identically affected by adaptation to a coloured light. So

8.5 Is there a special 'yellow' receptor? This experiment gives the answer. It uses an *anomaloscope* – an instrument giving a red + green mixture field (appearing yellow), next to a monochromatic yellow field appearing identical. Adaptation to a red or a green light does not produce a breakdown of the match between these two fields; from which it follows that there cannot be a separate mechanism for seeing yellow – it is always seen by the combined activity of the red and green receptor systems.

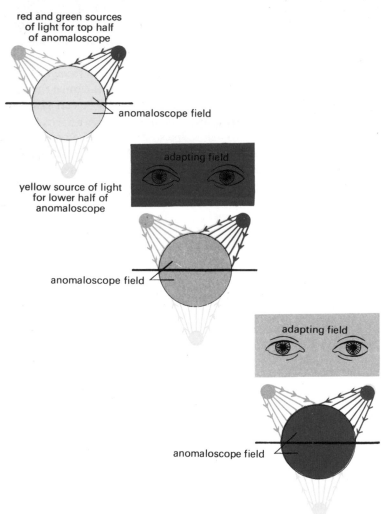

red and green sources of light for top half of anomaloscope

anomaloscope field

adapting field

yellow source of light for lower half of anomaloscope

anomaloscope field

adapting field

anomaloscope field

there is no special yellow receptor. The experiment may be repeated for other colours, with a similar result, showing that no colour as seen has a special system.

The same result is also obtained for anomalous observers: their initial setting is different, but it also remains unaffected by adaptation.

Now we come to a curious conclusion. If the anomaloscope cannot differentiate between the normal eye with and without colour adaptation, it follows that anomaly cannot be like colour adaptation. But this is exactly what colour anomaly *is generally supposed to be like*: namely a reduction in the sensitivity of one or more colour systems of the retina, through partial loss of a photo-pigment. This must be wrong. The reason for anomaly is not clear; there may be many causes, but it is certainly not due to a straightforward shortage of photopigment, or the anomaloscope would not work. Some kinds of anomaly may be due not to reduced sensitivity of colour response systems, as in colour adaptation, but to spectral shift of response curves. Others might be due to neural 'short circuiting' of colour receptor systems, so that two systems signal as though they were a single system.

9.1 René Descartes (1596–1650), perhaps the most influential of modern philosophers. It is now difficult to escape from his duality of mind and matter, which permeates all modern thought in psychology. He clearly described perceptual size and shape constancy, long before they were studied experimentally.

9 Illusions

Perception can go wrong in many ways. Most dramatic – an entire world may be created and mistaken for reality. This can happen in drug-induced states, or in mental disease. In addition to hallucinations, where experience departs altogether from reality, normal people may perceive surrounding objects in a distorted way. In this chapter we will pass rapidly over hallucinations, but spend some time on distortions giving rise to illusions of various kinds.

Hallucinations and dreams

Hallucinations are similar to dreams. They may be visual or auditory, or may involve any of the other senses such as smell or touch. They may even combine several senses at once, when the impression of reality can be overwhelming. Hallucinations can be socially determined, and there are cases of many people 'witnessing' together events which never occurred.

There are two ways of regarding hallucinations, and these two ways go deep into the history of thought. Dreams and hallucinations have always excited wonder, and sometimes more, for they have affected personal and political decisions, sometimes with bizarre and terrible results.

To the mystic, dreams and hallucinations are insights into another world of reality and truth. To these thinkers the brain is a hindrance to understanding – a filter between us and a supraphysical reality, which allows us to see this reality clearly only when its normal function is disturbed by drugs, disease, or starvation. To the more down-to-earth, however, including the empiricist philosophers, the brain is to be trusted only in health, and hallucinations although interesting and perhaps suggestive are no more than aberrant outputs of the brain, to be mistrusted and feared. Aldous Huxley, in his *Doors of Perception*, represents and describes most vividly the viewpoint of the mystic, but the majority of neurologists and philosophers hold that truth is to be found through the physical senses, while a disturbed brain is unreliable and not to be trusted.

To physiologists, hallucination and dreams are due to spontaneous activity of the brain, unchecked by sensory data. Although we have every reason to believe that this is indeed the cause, it runs counter to the classical empiricist's ideas of perception as being passive.

The Canadian brain surgeon, Wilder Penfield, produced hallucinations by stimulating the human brain with weak electric currents. Brain tumours may give persistent visual or auditory experiences, while the 'aura' preceding epileptic seizures may also be associated with hallucinations of various kinds. In these cases the perceptual system is moved to activity not by the normal signals from the sensory receptors, but by more central stimulation. Generally both the world as signalled by the senses and the strange hallucinogenetic world are experienced together. They may seem equally real, or the phantasy world may dominate to take over reality. It remains mysterious how hallucinogenic drugs, such as L.S.D. (d-Lysergic acid diethylamide) affect vision, though it is found that spontaneous activity of the lateral geniculate body is reduced. These drugs also affect animals: webspinning of the spider *Aranea diamata* is disrupted, and monkeys reach out for non-existent objects. Just as mysterious is why anaesthetics remove pain and all experience. Rather similar vivid imagery can occur in half-waking states (hypnagogic imagery) when the experience can be like looking at a technicolour film, with the most vivid scenes apparently passing before the eyes although they are shut.

Hallucinations have also been found to occur when people are isolated in solitary confinement, in prison or experimentally in isolation chambers in which the light is kept subdued or diffused with special goggles, and nothing happens for hours or days on end. It seems that in the absence of sensory stimulation the brain can run wild and produce fantasies which may dominate. It is possible that this is part of what happens in schizophrenia, when the outside world makes little contact with the individual so that he is effectively isolated. These effects of isolation are interesting not only from the clinical point of view: they may present some hazard in normal life. Men may be effectively isolated for hours with very little to do in industrial situations, where responsibility is taken from the operator by automatic machines which need only be attended to occasionally; and in single-handed sailing and space flights prolonged isolation might occur. The hazards are sufficient reason for sending more than one man into space.

In the author's opinion, there is no evidence for the mystic's attitude to hallucinations, for although the experiences may be extremely vivid they probably never convey information of 'objective' validity. But uncontrolled brain activity may indicate something of hidden motives and fears and so be revealing.

Figures which disturb

There are some figures which are extremely disturbing to look at. These can be quite simple, generally consisting of repeated lines. The series of rays as in figure 9.2 or parallel lines as in figure 9.3 were first extensively studied by D. M. MacKay, who suggested that the visual system is upset by the redundancy of such patterns. The point is that given a small part of this figure, the rest can be specified by simply saying 'the rest is like what is given'. MacKay suggested that the visual system normally uses the redundancy of objects to save itself work in analysing information. The ray figure is such an extreme case of a redundant figure that the system is upset by it. It is not entirely clear why this should happen, and one can think of other figures apparently just as redundant which do not upset the system, but it is an interesting idea. The ray figure has a curious after-effect; when looked at for a few seconds, wavy lines appear. These are seen for a time when the gaze is transferred to a structureless field. It is a moot point whether the ray pattern produces these effects because of small eye movements shifting the repeated lines upon the retina, and so sending massive signals from the 'on' and 'off' retinal receptors. If this is the explanation, the effect may be similar to the disturbance of flickering light. However this may be, the visual system certainly is disturbed, and this effect should be considered where repeated patterns are used, say in architecture.

9.2 Ray figure studied by MacKay. Is it the redundancy of the figure which disturbs the brain? Or do the repeated lines stimulate movement detectors, with each movement of the eyes? If a screen is looked at after looking at this figure, there will be an after-effect like grains of rice in movement. This also occurs after watching movement, in the waterfall effect.

The patterns of curved lines could be moiré patterns from after-images.

9.3 Closely spaced parallel lines have similar effects to the converging 'MacKay rays'. Such figures are extensively used in Op Art, especially by Bridgit Riley, with stunning effect.

9.4 The Muller–Lyer, or *arrow illusion*.
The figure with the outgoing fins looks longer
than the figure with the ingoing fins. Why?

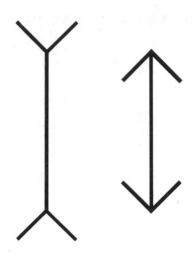

Visual distortions

Some simple figures are seen distorted. These distortions can be quite
large. Part of a figure may appear twenty per cent too long, or too
short; a straight line may be bent into a curve so that it is difficult to
believe it is really straight. Virtually all of us perceive these
distortions, and in the same directions for each figure.

Many theories have been put forward, but most can be refuted
experimentally or rejected as too vague to be helpful. We will have a
quick look at theories which can be safely rejected, before trying to
develop a more adequate theory. But first we should experience some
of the illusions themselves. Figures 9.4, 9.5 and 9.6 show some of the
best known illusions. They are given with the names of their
discoverers, mainly nineteenth-century physicists and psychologists
working in Germany, but it may be convenient to give some of them
descriptive names.

9.5 The Ponzo, or *railway lines illusion*. The upper horizontal line looks the longer. This same line continues to look longer in whichever orientation the figure is viewed. (Try rotating the book.)

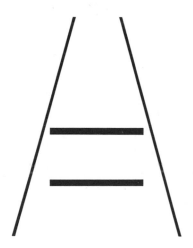

The best-known is the *Muller–Lyer* illusion, shown in figure 9.4. This is a pair of arrows whose shafts are of equal length, one having outgoing and the other ingoing arrow-heads at each end. The one with the outgoing heads looks considerably longer, though it is in fact the same length, as may be checked with a ruler. We may call this figure the *arrow* illusion (or if our theory, to be developed, is right, the *corner* illusion).

The second example is the *Ponzo illusion* (figure 9.5). The cross line in the narrower part of the space enclosed by the converging lines looks longer than the other cross line, although they are the same length.

Figure 9.6 shows four further distortion illusions: the Hering, Poggendorf, Orbison (two combined) and the Zöllner figures.

Psychologists and physiologists have tried to explain the distortion illusions for the last hundred years and any explanation is still controversial. I believe however that we can safely reject most of the

9.6 Four classical distortion illusions:
a Hering figure – red lines bowed;
b Poggendorf figure – red lines displaced sideways;
c Orbison figure (two combined) – red square
and circle distorted;
d Zöllner figure – red lines tilted.

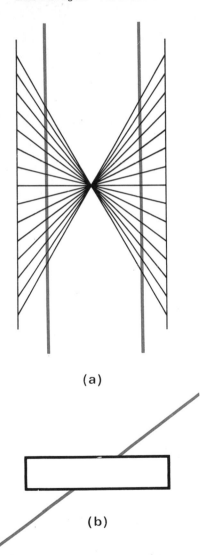

(a)

(b)

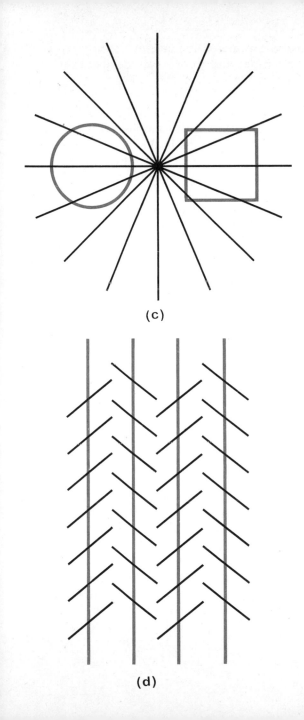

(c)

(d)

theories which have been proposed and develop a theory which both accounts for the distortions and throws considerable light on the nature of perception.

Theories we can reject

1 The eye movement theory. This theory supposes that the features giving the illusion make the eyes look in the 'wrong' place. In the arrow illusion, it is supposed that the eyes are drawn past the lines by the arrow-heads, which makes the lines look the wrong length, or – an alternative theory – that they are drawn within the lines. But neither of these can be correct. If the retinal image is optically fixed to the retina (figure 5.8); or more simply, as the reader can try for himself. by illuminating the figures with a photographic flash and viewing the after-image; the illusory distortions remain, undiminished. So eye movements cannot be held responsible.

The eye movement theory is sometimes stated in a rather different form, perhaps to avoid this difficulty. In this form it is not actual eye movements but the *tendency* to make eye movements which is supposed to produce the distortions. We can reject this with confidence, by the following consideration. The eyes can only move, or have a tendency to move, in one direction at a given time, but the distortions can occur in any number of directions at the same time. Consider the pair of arrows in figure 9.4. The first arrow is lengthened and the second shortened at the same time. How could this be due to an eye movement – or a tendency to an eye movement – which can occur in but one direction at a time? There is no evidence for the eye movement theories.

2 The limited acuity theory. Considering the arrow illusion – we should expect the figure with the outgoing fins to look too long, and the one with the ingoing fins to look too short if the acuity of the eye were so low that the corner could not be clearly seen. This may be demonstrated by placing a sheet of tracing paper over the figures, when a slight change of length might appear. The effect is however far too small to explain the Muller–Lyer illusion and it will not apply to many of the other illusions; so this is not a serious candidate.

3 Physiological 'confusion' theories. Although we believe that physiological processes mediate all perception, thinking and experience, we can have specific 'physiological' theories. They point to disturbance of components of mechanisms. Such disturbance might be caused by drugs, fatigue, or adaptation to intense or prolonged stimulation, as in the 'waterfall effect'. The distortion illusions are different from all these, for they occur in all normal people and immediately, without adaptation. Certain pattern features upset the perception of size and angle immediately. Could it be that these features directly upset the orientation detectors (pages 47 8) discovered by Hubel and Wiesel? We should distinguish two kinds of theory along these lines:

(i) That orientation detectors exaggerate all acute angles and minimise all obtuse angles. Although this was suggested by Helmholtz, and has been revived since, there is no convincing independent evidence. In any case, why are acute angles exaggerated? This occurs in some, though not all distortion illusions; but it is part of what needs to be explained.

(ii) That there are interactive effects between active orientation (or angle) detectors. It seems perhaps unlikely that the initial stages of pattern detection would be so 'ill-designed' that there would be serious interactions to generate errors, but a recent suggestion by Colin Blakemore has made this more plausible. Blakemore supposes that interactive effects capable of producing distortions may be side-effects of the process known as lateral inhibition. If the distortions were an unfortunate but inevitable consequence of such a process, then the apparent ill-design is understandable.

Lateral inhibition is interaction between regions, such that regions of strong stimulation reduce the sensitivity of surrounding regions. This gives a kind of 'sharpening' effect for gradients of neural stimulation, which is economical, for it is generally peaks of stimulation which are important. Blakemore's suggestion is that converging or angled lines produce asymmetrical regions of lateral inhibition, which will shift the neurally signalled peaks of stimulation, to produce visual distortions.

A somewhat related and sophisticated version of the lateral inhibition kind of theory has been advanced recently by A. Ginsberg, working in F. W. Campbell's laboratory at Cambridge, where ideas of pattern recognition by Fourier analysis are being developed. They suppose the eye to have 'spatial filters', which could distort certain shapes, rather as extreme audio filtering can distort sound.

We may call these *physiological* theories because they invoke disturbance in the information channels, or in functional units handling signals, rather than inappropriateness of how the signalled information is being applied to the perceptual situation. Cognitive theories, on the other hand, suggest that errors occur when knowledge, or strategies for seeing, are misapplied. Though so different, it is remarkably difficult to decide between 'physiological' and 'cognitive' theories of distortion illusions.

Here are some difficulties which arise if one supposes that the illusions have a 'physiological' origin:

(i) Shifts of position should be small, probably comparable to visual acuity, but the distortions are large.

(ii) It is the signalling of angles, rather than spatial positions or lengths, which are supposed to be disturbed. This raises the question: Are these illusions essentially distortions of angle, or of positions or lengths? Several illusions (e.g. the Muller–Lyer, the Ponzo, the Zöllner and the Poggendorf) would appear to be changes in lengths, or parallel displacements of lines, without change of angle.

(iii) The shifts in position can be cumulative across many parallel lines (as in the Zöllner illusion) but there seems no reason why shifts should be cumulative from one border to the next by lateral inhibition.

(iv) The illusions still occur though different parts are presented to each eye, to be combined by central stereoscopic fusion, by Julesz's technique (pages 69–72). The lateral inhibition could not then be retinal, but must occur after fusion of the two images, which is rather unlikely.

(v) The Muller–Lyer illusion occurs though the usual lines joining the arrows are left out. It persists even if the figure is reduced to three dots for each arrow. There seems no reason why lateral inhibition would occur to distort these figures for there is no relevant stimulation.

(vi) Distortions occur in several figures having only right-angular or parallel lines, when lateral inhibition should be symmetrical. Examples: the horizontal/vertical illusion (vertical lines being longer than horizontal); the Zöllner figure drawn with right-angular lines and omitting the parallels, when the 'herring bones' still appear

9.7 Muller-Lyer arrow heads (*left*), drawn as right-angles
and without shafts. Here there are only right-angles
and parallel lines, and yet the illusion still occurs.
This is difficult to explain on a 'lateral inhibition' theory;
but this is still a perspective figure, and could set
constancy size scaling to produce the distortion.

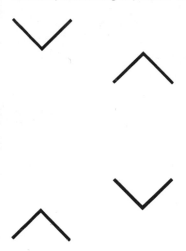

9.8 The *absence* of perspective generates distortion. The further edge would
normally be shrunk by distance – to be compensated by constancy scaling.
Here, evidently, the scaling is set by the assumption of depth, to produce
distortion. (Hide the legs, and the distortion will disappear. It must
be recognised as an object of familiar shape for distortion to occur; so this
is not quite the same as the Muller-Lyer illusion, though it is related.)

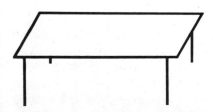

9.9 Man enough for the job? Perhaps we identify ourselves with pillars, so that there is a right size in human terms for carrying the load. Is this the basis for distortion illusions? (Caryatids of the Erechtheion, Athens, 421–405 BC, designed by Mnesicles).

displaced; the Muller–Lyer arrows with right-angular heads and the 'shafts' omitted, when the separation between the heads is still upset (figure 9.7); and the rectangular table figure (figure 9.8).

We will now look at what we regard as cognitive theories of these distortion illusions. We start with:

4 The empathy theory. This theory was suggested by Theodore Lipps, and is based on an idea of the American psychologist R. H. Woodworth. The idea is that the observer identifies himself with parts of the figure (or with say the pillars of a building) and that he becomes emotionally involved so that his vision is distorted rather as emotion may distort an intellectual judgment. In the arrow illusion, it would be argued that the outgoing arrow suggests, emotionally, expansion which one then sees.

It is true that a very thick column supporting a narrow cornice on a building looks clumsy; and perhaps one does imaginatively stand in the place of the column, much as Hercules took the load of the sky off the shoulders of Atlas before turning him to stone. The caryatids of Greek temples (figure 9.9) embody (quite literally) this idea in architecture. But although of immediate relevance to aesthetics, it can hardly be taken seriously as a theory of the illusions. The arrow figure (for example) gives distortion whatever one's mood, and continues to do so when any initial emotional response would surely have died through boredom. There may be perceptual effects of strong emotion, but the illusion figures would seem singularly devoid of emotional content – except to those who try to explain them! More serious: the distortions are virtually the same for all observers though emotions are very different.

5 The pregnance or 'good-figure' theory. The idea of 'pregnance' is central to the German Gestalt writers' account of perception. The English meaning is similar to the use of the word in 'a pregnant sentence'. A 'pregnant' figure is one expressing some characteristic although it is not all present. The illusions are supposed to be due to pregnance exaggerating the distance of features seeming to stand apart, and reducing the distance of the features which seem to belong together.

The status of the idea of pregnance is doubtful. Certainly random

9.10 An inside corner. The edges of the ceiling and floor form the Muller–Lyer figure. The inside corner recedes in depth.

9.11 An outside corner. Here the corner approaches in depth.
(A corner of the Cambridge Psychological Laboratory).

or systematic arrangements of dots do tend to be grouped in various ways, so that some belong to one figure while others are rejected or are organised into other patterns (figure 1.1) but there seems no tendency for the dots to change their position as a result of such grouping; and surely this would be an inevitable prediction of the pregnance theory of distortion.

6 The perspective theory. This theory has a long history, into which we need not enter, but the central idea is that the illusion figures suggest depth by perspective, and that this suggestion of depth produces size changes.

Consider the two illusion figures we started with (figures 9.4 and 9.5). Each can be fitted very naturally to typical perspective views of regular shaped objects lying in three dimensions. The illusion figures can be thought of as flat projections of three dimensional space – simple perspective drawings – and the following generalisation holds:

150

Those parts of illusion figures which would represent distant features are enlarged, and those parts corresponding to near features are shrunk.

This is seen clearly in the Muller–Lyer illusion. The arrow with the outgoing fins could represent the inside corner of a room (figure 9.10). The ingoing arrow-heads could represent an outside corner of a building (figure 9.11). The railway (figure 9.12) is an explicit perspective picture – placing the upper rectangle further than the lower, and expanding it as in the Ponzo line figure (figure 9.5).

It should be made clear at once, however, that although the illusion figures do seem to be typical flat projections of three dimensions, each one *could* always represent something quite different. The arrow figures could represent a steeplejack's view of a roof: the converging lines of the railway tracks illusion *could* simply be a pair of converging lines, rather than parallel lines seen as converging because of distance. The illusion figures are typical perspective views; but in all cases they could be drawings of something quite different.

The traditional perspective theory simply states that these figures 'suggest' depth, and that if this suggestion is 'followed up' the most distant features represented appear larger. But how could 'suggestions of distance' produce changes of apparent size? This is too like the empathy theory. Further, why should suggestion of *greater* distance produce *increase* in size – when distant objects are normally seen somewhat smaller when further away? The traditional perspective theory fails to provide a *modus operandi* for the distortions; and its predicted size changes are the wrong way round.

Although the predictions of the perspective theory go exactly the wrong way, this is far better than predictions which are quite unrelated to the facts. In mathematics or logical theories it is very easy to get a sign wrong: has something like this happened here? We will try to develop a theory along these lines which leads to correct predictions, as well as linking the illusions to other perceptual phenomena. It is worth devoting some time to this, for it is by

9.13 Size constancy. The image of an object halves in size with each doubling of the distance of the object. But it does not *appear* to shrink so much. The brain compensates for the shrinkage of the image with distance, by a process we call *constancy scaling*.

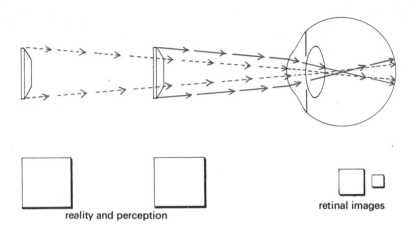

reality and perception

retinal images

establishing connections between phenomena that we gain understanding. Illusions become no longer trivial effects of certain patterns, but rather tools for investigating basic processes involved in perceiving the world.

There is a perceptual phenomenon which is quite capable of producing distortions – Size Constancy. This is the tendency for the perceptual system to compensate for changes in the retinal image with changes of viewing distance. It is a remarkable and fascinating effect which we can see operating in ourselves. It can go wrong, and when it does, instead of keeping the scale of things constant in spite of changes of retinal image size with changes of object distance – it produces distortions. The corrections necessary for normally keeping objects nearly constant in apparent size are large – so are illusions.

The image of an object doubles in size whenever its distance is halved. This is a simple fact from geometrical optics, and applies to a camera as it does to the eyes. Why it occurs should be clear from figure 9.13. Now what is odd, and it certainly requires explanation, is the fact that although the *image grows* as the distance of the object decreases, *it still looks almost the same size*. Consider an audience at a theatre – the faces all look much the same size, and yet the retinal images of the distant faces are far smaller than the nearer. Look at your two hands, one placed at arm's length the other at half the distance – they will look almost exactly the same size, and yet the image of the further hand will be only half the (linear) size of the nearer. If the nearer hand is brought to overlap the further, then they *will* look quite different in size. This little experiment is well worth carrying out. The overlap defeats Constancy Scaling, and shows us what perception would be like without it.

What is known as Size Constancy was described by Descartes, who wrote in his *Dioptrics*, of 1637:

I need not, in conclusion, say anything special about the way we see the size and shape of objects; it is completely determined by the way we see the distance and position of their parts. Thus, their size is judged according to our knowledge or opinion as to their distance, in conjunction with the size of the images that they impress on the back of the eye. It is not the absolute size of the images that counts. Clearly they are a hundred times bigger [in area] when the objects are very close to us than when they are ten times farther away; but they do not make us see the objects a hundred times bigger; on the contrary, they seem almost the same size, at any rate so long as we are not deceived by (too great) a distance.

We have here as clear a statement of size constancy as any made later by psychologists; except that Descartes over-states the importance of knowing the sizes or shapes of the objects. Descartes goes on to describe what is now called Shape Constancy:

Again, our judgments of shape clearly come from our knowledge, or opinion, as to the position of the various parts of the objects and not in accordance with the pictures in the eye; for these pictures normally contain ovals and diamonds when they cause us to see circles and squares.

The ability of the perceptual system to compensate for changing distance has been very fully investigated, notably by the English psychologist Robert Thouless in the 1930's. Thouless measured the amount of constancy under various conditions, and for different types of people. He used very simple apparatus – nothing more elaborate than rulers and pieces of cardboard. For measuring size constancy he placed a square of cardboard (which, unlike say a hand, could be any size) at a given distance from the observer, and a series of different sized squares at a nearer position. The subjects were asked to choose the nearer square which appeared the same size as the further square. From the actual sizes (measured by ruler) the amount of size constancy could be calculated. Thouless found that his subjects generally chose a size of square almost the same as the actual size of the distant square, although its image was smaller than the image of the nearer square. Constancy was generally almost perfect for fairly near objects, though it broke down for distant objects, which do look small, like toys. Thouless also measured shape constancy, and this he did by cutting out a series of cardboard 'lozenges', or ellipses of various eccentricity, which were selected by the subject to match for shape a cardboard square, or circle, placed at an angle to the subject's line of sight – the comparison lozenge or ellipse being placed normal to the subject's line of sight. Again it was found that constancy could be nearly perfect, depending largely on the depth cues available. Thouless described this as 'phenomenal regression to the real object'. But this might be taken to imply that we have some *other* knowledge of object reality, so this phrase is no longer used. There has been serious confusion over Constancy as a *process* and Constancy as the *result* of (brain) processes. The maintenance of apparent size and shape is clearly the *result* of processes. We may think of this as the brain *scaling* the retinal image, to represent the sizes of objects lying at various distances though their images shrink with increasing distance of the object.

It is possible, with a simple demonstration, to see one mechanism of one's own size scaling at work.

Emmert's Law. First, obtain a good clear after-image, by looking steadily at a bright light, or preferably a photographic flash. Then look at a distant screen or wall. The after-image will appear to lie on the screen. Then look at a nearer screen. The after-image will now appear correspondingly nearer – and will look smaller. With a hand-held screen, (such as a book) moved backwards and forwards, the after-images will expand as the hand moves away and shrink as it approaches the eyes: though of course the retinal after-image remains of constant size. We therefore see the brain's *scaling* changing as the

9.14 The Muller-Lyer illusion destroyed. When the stereo depth and the perspective are appropriate, at the correct viewing distance (40 cm) there is no distortion. This result is predicted from the Inappropriate Scaling Theory of the distortion illusions. (Each measured point is the average of sixty readings. The vertical bars represent standard errors Gregory and Harris, 1975.)

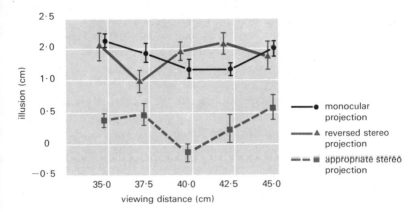

distance of the screen changes. The after-image is seen to (nearly) double in size with each doubling in distance of the screen. This is Emmert's Law.

Clearly we are seeing size changes which would normally compensate changes of the retinal images with distance. Here, for fixed after-images, the retinal image is fixed in size though its apparent distance is changed: so we see Constancy Scaling at work – producing changes of perceived size.

For any object there is only one correct perceived size. This is normally given when the distance is correctly perceived. Can scaling be set inappropriately – to produce the distortion illusions? Is this the key to an adequate theory of distortion illusions?

The Inappropriate Constancy-Scaling Theory It is one thing to have a theory and quite another to prove it; or to show that it is the best of available suggestions. If indeed these distortion illusions are always related to distance, then it may indeed seem reasonable to think that perspective sets Size Scaling, to give errors of perceived size when

155

distance is seen incorrectly. This surely looks reasonable: but there is a difficulty, which blocked development of a theory of illusions along these lines for many years. The difficulty is that the illusion figures *generally appear flat.* If constancy always follows perceived depth, then obviously there is something seriously wrong, or something missing from this theory. We shall take the view that the theory is essentially correct but not yet complete. Can we complete it?

The reason why the illusion figures appear flat in spite of their perspective features is not difficult to discover. In the first place, if they are shown on normal textured paper, there is clear visual evidence that they are indeed flat. Perspective pictures have a kind of depth paradox. They *depict* depth in their perspective and other 'depth cues', and yet as objects they *are* flat, and their surfaces provide depth cues of their true flatness. We can however remove these surface cues – by making the surface invisible. This can be done by painting or drawing with luminous paint. The illusion figures (or any perspective drawings) appear in dramatic depth when the background is removed in this way. This is very well worth trying. Depth can still however be lost, or reduced, when the perspective angles are very different from the angles which would be given from a reasonable viewing distance. There is strictly only one position from which a perspective picture gives correct perspective. We have some tolerance for error here; but the illusion figures are often drawn with highly exaggerated perspective – strictly requiring an impossible viewing distance, of a few centimetres from the figure. For realistic depth to be given by the luminous figures moderate angles should be used and they should be viewed from the appropriate distance.

We can get over the difficulty that the distortions occur though the illusion figures appear flat by supposing that *perspective can set Size Scaling directly.* Although there is nothing implausible in this suggestion, it does flout what has until recently been assumed about Constancy – that it always follows seen depth. This has been held with great authority, as is clear from the following quotation from W. H. Ittelson, who cites support from five other distinguished psychologists who have worked on this problem. He has this to say: 'Constancy, it is universally agreed, is dependent upon the proper estimation of distance.' Possibly the Emmert's Law demonstration with after-images (page 154) may suggest that size change always follows apparent distance: but neither it nor anything else proves it; so we may challenge this assumption. We suggest that although scaling can indeed follow perceived distance (as in the Emmert's Law demonstration with after-images) it can *also* be set directly by depth cues, especially by perspective. We see this when perceived depth is countermanded by the background texture of figures or pictures. We should then expect distortion illusions to occur when perceived distance is incorrect, and also when there are misleading depth cues.

We suggest, then, that there are two kinds of scaling: 'upwards' from depth cues and 'downwards' from perceived depth.

Measuring distortion and depth

It is quite easy to measure the amount of an illusion of the kind we are considering here – distortions of size or shape. It can be done by showing a set of comparison lines, or shapes, and asking the observer to select the one most like the illusion figure as he sees it. But of course it is essential to show the comparison line in such a way that *it* is not distorted. It is generally best to arrange for the comparison line to be continuously adjusted, either by the observer or by the experimenter. A suitable apparatus is shown in figure 9.15.

Measuring apparent depth may seem to be impossible. But consider figure 9.16. The figure is presented back-illuminated, to avoid texture, and it is viewed through a sheet of polaroid. A second sheet of polaroid is placed over one eye, crossed with the first so that virtually no light from the figure reaches this eye. Between the eyes and the figure is a half-silvered mirror, through which the figure is seen and which reflects one or more small light sources adjustable in distance.

These appear to lie in the figure: indeed optically they *do* lie in the figure, provided the path length of the adjustable light (or lights) is the same as the distance of the figure from the eyes. But the small reference lights are seen with *both* eyes, while the figure is seen with only *one* eye – because of the crossed polaroid filters. By adjusting the reference lights in distance, they may be placed at the same *apparent* distances as any selected features or regions of the figure. If the figure has perspective, or other depth cues, then the lights are placed by the observer not at the true distance but at the apparent distances of the

9.15 How to measure an illusion. The observer sees a single arrow figure and an adjustable comparison line, which is set to appear the same length as the distorted line. This gives a direct measure of the extent of the illusion. (Measurement is only possible, however, when the illusion is not logically paradoxical.) The figure shows a back view of the apparatus.

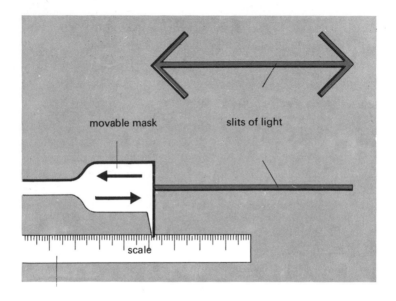

movable mask   slits of light

scale

features he is matching for distance as he sees it. For people with normal stereoscopic depth perception this is a quite easy task and gives consistent measures of apparent distances.

The technique shows that the illusion figures are indeed seen in depth, according to their perspective features – when this depth information is not countermanded by visible texture of the picture plane. It is found that increased *distance* of features correlates highly with illusory *expansion*. So we have objectively related these illusory distortions to perspective depth.

The finding that measured apparent depth is closely related to these measured distortions is suggestive; but it is hardly proof that perspective depth features produce the illusory expansion. Can we think of a more direct test to distinguish between distortions due

9.16 How to measure subjective visual depth. The (flat) figure is back illuminated, to avoid texture which gives it paradoxical depth. The light from the figure is cross-polarised to one eye. An adjustable reference light is introduced into the figure, by reflection from a half-silvered mirror. This is seen with *both* eyes, and is set to the apparent distance of any selected parts of the figure. Thus binocular vision is used to measure monocular depth.

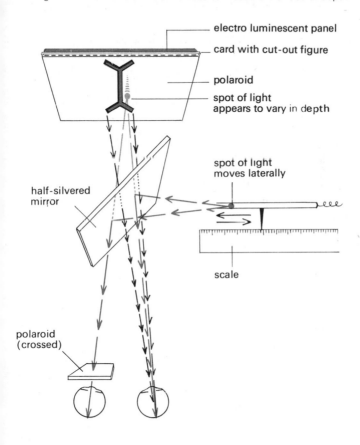

electro luminescent panel

card with cut-out figure

polaroid

spot of light
appears to vary in depth

spot of light
moves laterally

half-silvered
mirror

scale

polaroid
(crossed)

to errors of the *signals* of the visual channel and errors due to *information* (such as perspective) applied inappropriately? We may have this in a recent experiment carried out by the author with John Harris, in which the distortions were found to vanish for these figures, when the geometry of the viewing conditions match the normal conditions for seeing typical objects in three dimensional space.

Illusion-destruction If the distortions of these figures are due to inappropriate Constancy Scaling – then the distortions should no longer occur when all the features setting the scaling are *appropriate*. To make them appropriate, we ensure that the perspective of the figures is exactly the same at the eyes as for viewing an object (such as corners or receding parallel lines) directly. We further ensure that it is seen in its correct depth – as though it were an object although it is in fact a flat picture. Consider the Muller–Lyer illusion figure: this is a flat perspective projection of corners. We make three dimensional wire model corners, and project them in accurate perspective, by casting their shadows on a screen with a small bright light source. When this is viewed with the eyes at the same distance as the light source, the perspective of the retinal image is exactly correct – it is the same as when viewing the object directly. But it will still appear flat. To give the correct three dimensional perception wc add a second shadow-casting light, which is separated horizontally from the first, by the interocular separation of the observer's eyes. Finally, it is so arranged that one eye sees one of the perspective shadow projections, and the other eye sees only the other projection, by cross polarization. The projections are combined by the brain – to give a three dimensional perception, exactly as though it were a real corner although it is flat on the screen. In this situation, the distortion is entirely absent. It is absent although the visual system is signalling the same angles as the usual illusion figure appearing flat and distorted: – so how could the illusion be due to distortions of signals of the visual channel? This seems to be clear evidence that the distortions are produced by depth cues when they are inappropriate to the depth of the object, and also when distances are seen incorrectly. The first kind of error is incorrect scaling 'upwards' from the depth cues: the second is incorrect scaling 'downwards' from perceived depth. This distinction is discussed more fully in the author's *The Intelligent Eye* (1970).

160

9.17 The circular culture of the Zulus. They experience few straight lines or corners and are not affected by the illusion figures to the same extent as people brought up in a 'rectangular' Western culture.

Cultural differences

In the Western world rooms are nearly always rectangular; and many objects, such as boxes, have right-angled corners. Again, many things, such as roads and railways, present long parallel lines converging by perspective. People living in the Western world have a visual environment rich in perspective cues to distance. We may ask whether people living in other environments where there are few right angles and few long parallel lines, are subject to the illusions which we believe to be associated with perspective. Fortunately, several studies

161

have been made on the perception of people living in such environments, and measurements have been made of their susceptibility to illusion figures.

The people who stand out as living in a non-perspective world are the Zulus. Their world has been described as a 'circular culture' – their huts are round, they do not plough their land in straight furrows but in curves, and few of their possessions have corners or straight lines. They are thus ideal subjects for our purpose. It is found that they experience the Muller-Lyer arrow illusion to only a small extent, and are hardly affected at all by other distortion illusion figures.

Studies of people living in dense forest have been made. Such people are interesting in that they do not experience distant objects, because they live in small clearances in the forest. When they are taken out of their forest, and shown distant objects, they see these not as distant, but as small. People living in Western cultures experience a similar distortion when looking down from a height. From a high window objects look too small, though steeplejacks and men who work on the scaffolding and girder structure of skyscrapers are reported to see objects below them without distortion.

By using the depth measuring technique (figure 9.15) Jan Deregowski has found that Zulus showing lack of illusion also show little or no perceived depth in these figures, though they are presented in conditions which do give depth by perspective for Western observers. So there now seems to be clear evidence for cultural factors in these illusions related to distance cues available in the environment.

There are illusions similar to those of vision found also for touch. This presents something of a problem for the theory we have been advocating. John Frisby has recently provided evidence that people with vivid visual imagery tend to have greater than usual touch illusions. Possibly touch information is interpreted according to visual processing and visual 'models' of the world – and distorted by visual scaling processes.

10 Art and reality

Perspective as we know it in Western art is extraordinarily recent. In all known primitive art, and in the art of all previous civilisations, there is no perspective until the Italian Renaissance. In the highly developed formalised painting of the ancient Egyptians, heads and feet are shown in profile, never foreshortened by perspective; which gives the figures a certain resemblance to child art. Chinese drawing and painting is most curious in this respect, for distance is represented by formal rules which contravene geometry, and which often give what we would regard as reversed perspective – lines diverging rather than converging with increasing distance. It is an extraordinary fact that simple geometrical perspective took so long to develop – far longer than fire or the wheel – and yet in a sense it has always been present for the seeing. But is perspective present in nature? Is perspective a discovery, or an invention of the Renaissance artists?

The laws and principles of perspective were first clearly described by Leonardo da Vinci (1452–1519) in his *Notebooks*, where he outlines a suitable course of study for the artist including, as well as perspective, the arrangement of surface muscles, the structure of the eyes of man and animals, and botany. He called perspective 'the bridle and rudder of painting', describing it in the following way:

Perspective is nothing else than the seeing of a plane behind a sheet of glass, smooth and quite transparent, on the surface of which all the things approach the point of the eye in pyramids, and these pyramids are intersected on the glass plane.

Leonardo treated the perspective of drawings as a branch of geometry. He described how perspective could actually be drawn directly on a sheet of glass; a technique used by the Dutch masters and, in a later form, with the *camera obscura* which employs a lens to form an image of the scene which may be traced directly. The projection is determined simply by the geometry of the situation and this constitutes so-called *geometrical perspective*; but as Leonardo

10.1 An Egyptian scene. The figures are shown in characteristic positions but without perspective. Perspective was not introduced into art until the Italian Renaissance.

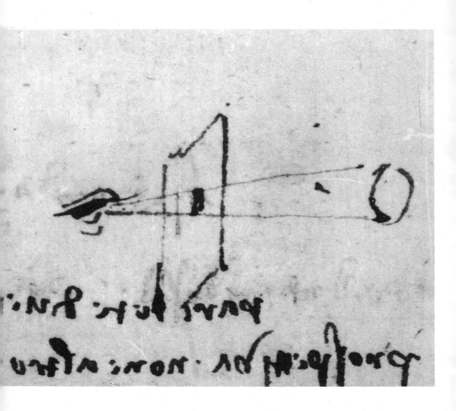

realised more clearly than many later writers, there is more to the matter than the pure geometry of the situation. Leonardo included in his account of perspective such effects as increasing haze and blueness with increasing distance; the importance of shadows and shading in drawings to represent the orientation of objects. These considerations go beyond pure geometry.

Consider a simple ellipse, such as in figure 10.6. This might represent an elliptical object seen normally or a circular object seen

10.3 This Canaletto is a fine example of perspective. It is worth pondering: has he painted the geometrical perspective, as given in the image in his eye, or has he painted the scene as he saw it – after his size constancy scaling has compensated for shrinkage of the image with distance?

10.4 Chinese 'perspective'. This is very odd, for it is neither geometrical nor as the world appears through constancy scaling. Presumably the Chinese adopted highly conventionalised symbolic representation.
10.5 (Right) An early example of perspective: *The Annunciation* by Crivelli (*c*. 1430–95).

OPVS CARO
LI·CRIVELLI
VENETI

LIBERTAS ☩ ECCLESIASTICA

10.6 and **10.7** Is the shape below in perspective?
Only if we know what it is. In the cartoon (right)
the same shape is clearly a circular object tilted.

obliquely. This figure does not uniquely indicate any one kind of object; it could be a projection of any of an infinite variety of objects, each seen from a certain angle of view. The art of the draughtsman and painter is, in large part, to make us accept just one out of the infinite set of possible interpretations of a figure, to make us see a certain shape from a certain point of view. This is where geometry goes out and perception comes in. To limit the ambiguity of perspective, the artist must make use of perceptual distance cues available to a single eye. He is forbidden the binocular distance cues of convergence and disparity, and also motion parallax. Indeed these cues will work against him. Paintings are generally more compelling in depth when viewed with a single eye, with the head kept still.

We have to consider a double reality. The painting is itself a physical object, and our eyes will see it as such, flat on the wall, but it can also evoke quite other objects – people, ships, buildings – lying in space. It is the artist's task to make us reject the first reality while conveying the second, so that we see his world and not mere patches of colour on a flat surface.

As we have seen from the example of the ellipse, a picture can

10.8 and **10.9** The puddle below is clearly lying flat on the ground — this is what puddles do. (Right) The same shape as the puddle in the cartoon, but how does it lie in space? It could be upright.

represent a given object from one viewing position, or any of an infinite set of somewhat different objects seen in some other orientations. This means that for the picture to represent something unambiguously, we must see what the object really is – what its shape is – and how it lies in space. It is very much easier to represent familiar than unfamiliar objects. When we know what the object is, then we know how it must be lying to give the projection given by the artist. For example, if we know that the ellipse is representing a circular object, then we know that this must be lying at a certain oblique angle – the angle giving the eccentricity drawn on the flat plane by the artist. We all know that wheels, dinner plates, the pupil of the human eye, etc., are circular objects, and for such familiar objects the artist's task is easy. We may see how easy it is from the power that very simple line drawings have for indicating form, orientation and distance – when the object is familiar. Consider the drawing of the boy with the hoop in the cartoon (figure 10.7). It is quite clear that the ellipse represents a circle at an oblique angle, because we know that it represents a hoop, and we know that hoops are circular. The hoop in this figure is in fact the same as the ellipse seen without context in figure 10.6. But now we know what it is we know how to see it. It would have been far more difficult for the artist to represent a distorted hoop.

Look at the amoeboid shape of the spilled wine in figure 10.8. It is seen as lying on a flat surface (the road), although the shape alone could equally well represent an infinity of shapes lying in various orientations. Suppose we remove the rest of the drawing, so that we have no clue as to what it represents, figure 10.9 shows just the puddle. It could equally well be something of rather indeterminate shape standing up and facing us. (Does it not look slightly higher in

the full drawing where it is clearly a puddle lying on the ground, than when it is an indeterminate shape? Does the context provide constancy scaling?) Although the figure is so simple it is evoking vast experience of objects – such as what happens when we drop bottles – and this physical knowledge determines how we see the amoeboid shape.

We may now take another example, again of an ellipse in a cartoon, but this one illustrates a rather different point. Take the ellipse shape in figure 10.10. This is of some interest, for it is presented without any perspective, and yet it clearly lies on the floor. It is seen as a circle. The child below (not shown yet seen!) could be cutting an elliptical hole, but we assume he is cutting a circle. This puts our viewing position at a certain height above ground, which is not determined by any other feature in the drawing but only by our interpretation of the meaning of the shape, based on our knowledge of small boys.

When an artist employs geometrical perspective he does not draw what he sees – he represents his retinal image. As we know, these are very different, for what is seen is affected by constancy scaling. A photograph represents the retinal image – not how the scene appears. By comparing a drawing with a photograph taken from exactly the same position, we could determine just how far the artist adopts perspective and how far he draws the world as he sees it after his retinal images are scaled for constancy. In general, distant objects look too small in a photograph – it is a common and sad experience that a grand mountain range comes out like a pitiful row of mole hills.

The situation here is curious. The camera gives true geometrical perspective, but because we do not see the world as it is projected on the retina, or a camera, the photograph looks wrong. It should not surprise us that primitive people make little or nothing of photographs. Indeed it is fortunate that perspective was invented before the photographic camera: we might have had difficulty accepting photographs as other than weird distortions. As it is, photographs can look quite wrong, particularly when the camera is not held horizontally. Aiming a camera upwards, to take in a tall building, gives the impression of the building falling backwards. And yet this is the true perspective. Skyscrapers do look slightly converging, though not so much as in a photograph taken from the same position and with the camera tilted up at the same angle as the eyes. Some

10.10 Another ellipse. This time we assume it is a circle, and see it as flat, because we know that the boy under the floor (almost seen!) would generally saw a circular hole.

architects have recognised that the visual compensation for distance is less efficient when looking upwards, and have built their towers to diverge slightly from the bottom to the top. The most notable example is the magnificent Campanile at Florence, designed by Giotto. Here the artist as architect has applied reversed perspective to reality, to compensate for the eye's inadequacy in correcting for perspective. There are examples of this on the horizontal plane also, notably the Piazza San Marco in Venice which is not a true rectangle but diverges towards the cathedral, so that it appears to be a true rectangle when the cathedral is viewed from across the Piazza. We find similar 'distortions' of reality to suit the eye and brain in Greek temples; though the Greeks never discovered the projective geometry of perspective.

We begin to see why it took so long for perspective to be adopted by painters. In an important sense perspective representations of three dimensions are wrong, for they do not depict the world as it is seen but rather the (idealised) images on the retina. But we do not see our retinal images; and we do not see the world according to the size or shapes of the retinal images, for these are effectively modified by constancy scaling. Should not the artist ignore perspective and draw the world as he sees it?

If the artist ignores perspective altogether his painting or drawing will look flat, unless indeed he can utilise other cues to distance with sufficient force. This seems to be almost impossible. If he did succeed in suggesting depth by other means, then the picture would look wrong, for these other cues would trigger the constancy scaling system to expand the more distant objects as represented. This means that the artist should use perspective – draw distant objects smaller – if the viewer's constancy scaling is affected by the depth cues he provides. Indeed, if he could provide *all* the normal depth cues, he should use *complete* perspective: so that the viewer sees sizes and distances as though he were seeing the original three dimensional scene. But – and this is the important point – in fact the artist can hardly hope to provide all the depth cues present in reality and so he should use a modified perspective for maximum realism.

The Ames perspective demonstrations

An American psychologist, Adelbert Ames, who started life as a painter, produced a series of most ingenious and striking perceptual demonstrations. Most famous is his Distorted Room. The further wall is sloped back at one side, so that it does not lie normal to the observer; but perspective is used to make this oddly shaped room give the same retinal image as a normal rectangular room. Now just as there is always an infinite set of arrangements of objects and orientations which could give a particular retinal image, so there is an infinite set of distorted rooms which could give the same images as those of the normal rectangular room.

What does an Ames distorted room look like? It looks like a normal rectangular room! There is really nothing surprising in this: it *must* look like a normal room if constructed according to strict perspective, and viewed from the right position – because the image it gives is the same as for an ordinary room. But if now we place objects in the room, very odd things happen. An object placed at the further corner shrinks. It looks too small because the image is smaller than would be expected for the apparent distance of that part of the room. In this way an adult may be shrunk to appear smaller than a child (figure 10.11). It is important that this effect still works in the photograph. In fact one does not really need the room to get the effect, for the photograph gives the same retinal image as the room, except that the photograph has, in addition, the flat-plane texture of the paper on which it is printed.

Evidently we are so used to rectangular rooms that we accept it as axiomatic that it is the objects inside (the two figures) which are odd sizes, rather than that the room is an odd shape. But this is essentially a betting situation – it *could* be either, or both which are peculiar. Here the brain makes the wrong bet, for the experimenter has rigged the odds. The interesting feature of the Ames Distorted Room is its implication that perception is a matter of making the best bet on the available evidence. It has been reported that wives may not see their husbands distorted by the Room – they see their husbands as normal, and the room its true queer shape. Behold the power of love?

To recapitulate: the empty Ames Distorted Room tells us nothing about perception. If properly constructed, it must look like a normal

10.11 Impossible? We accept that the room is rectangular though in fact it is not, and see the figures as different sizes. This is what happens inside the Ames distorted room. We are so used to rooms being rectangular that we bet on the room being normally shaped. Here we are wrong.

10.12 The geometry of the Ames distorted room. The further wall in fact recedes from the observer (and the camera) to the left. The figure on the left is further away, but the walls and windows are arranged to give the same retinal image as a normal rectangular room, and the figures appear the same distance and different sizes. (The nearer figure is about doubled in size in this room.)

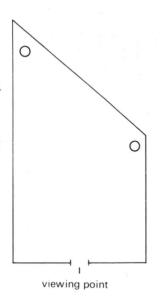

viewing point

rectangular room giving the same projection from the observer's point of view. It must also look the same to the camera, or to any conceivable optical device or other kind of eye which does not get information of distance by other means. But when objects (such as people) are added, the room brings out the point that perceptual interpretation involves betting on the odds. This shape is so unlikely for a room (at least for Western eyes) that our perception goes wrong – when the truth is so improbable. This tells us something about the importance of previous experience and learning in perception. It is only very familiar objects (husbands) which refuse to be distorted by the room. Familiarity with the room, especially through touching its walls, even with a stick held in the hand, does gradually reduce its

179

10.13 A surface texture represented by perspective. The surface could be a random texture. Texture is the visible surface structure of objects, or of the ground between objects. It is independent of viewing position. Perspective is to be found only in images and pictures – not in objects, or 'reality'.

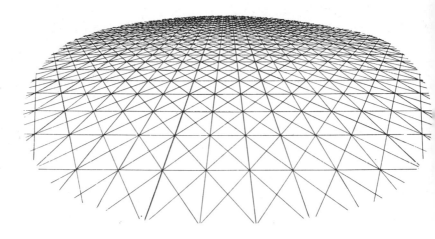

distorting effect on other objects, and finally it comes to look more or less its true odd shape.

Another celebrated Ames demonstration is the Rotating Window. This is a flat, non-rectangular, window-like object made to rotate slowly by means of a small motor. It has shadows painted upon it (so that as it rotates the light source apparently casting the painted shadows would have to rotate, improbably, with it since the shadows on the 'window' never change in length). What is seen is a complex series of illusions. The direction of rotation is ambiguous, seeming to change spontaneously. (This is the 'windmill' effect, observed when the rotating vanes are seen against the sky. The direction of rotation reverses spontaneously while one looks at it.) Any small object attached to the rotating window will suddenly seem to move in the wrong direction, when its movement is seen correctly though the window in rotation is seen falsely.

The window may change dramatically in size – a remarkable and striking effect. Evidently constancy has been upset by the atypical transformations at the eye of this unlikely window. The demonstration is very dramatic but too complicated to be a good research tool.

Gibson's gradients

The experimental work of J. J. Gibson, at Cornell, is justly celebrated. Gibson emphasizes the importance of texture gradients, and of motion parallax from the observer's movements. He considers that perceived depth and form are determined by these, and other features of what he calls the 'ambient optical array' – the surrounding pattern of light. Perception, he believes, is somehow given to us *directly*, from the 'ambient array' of light, even without the need for retinal images. He thus goes back to pre-tenth century ideas, before Alhazen's discovery of the principle of the *camera obscura*. Gibson's theory is a kind of 'Realism', in which perceptions are supposed to be *selections*, rather than representations, of the object world. Realism is popular among philosophers as it seems to give certainty to perception, and so to give an undoubted basis for empirical knowledge. This is not to be expected of perception regarded as *representations*, which is how we regard perception in this book. Gibson's attempt at a Realist theory of perception is brave and interesting: but how can it be tenable, from all we know of eyes and images and the unreliability of sensory signals?

However this may be, Gibson has taught us a lot about depth cues (though he might not like this term) and is undoubtedly right to get away from thinking of perception merely in terms of a stationary observer viewing with a single eye.

Texture gradients, (figure 10.13); hiding parts of far objects by nearer objects, (figures 10.14 and 10.15) – as well as motion parallax, stereopsis, haze associated with distance and many more – are all features signalling depth and distance. On a Realist theory, colours and shapes are 'selected' passively, and become literally part of us. On a representational theory they are 'data' or 'cues' used for inferring depth and form by active brain processes. Specific evidence for this is that none of these features *determine* depth or form; and they may be pitted against each other to produce ambiguities, paradoxes and distortions – which cannot belong to the world.

10.14 A cunning trick which upsets depth. The two sets of squares, though they look the same, are differently arranged in distance. In the third row of playing cards, the jack is in fact nearer than the six.

The power of the various 'depth cues' can be measured with the technique described in the previous chapter (page 157). No features *determine* depth or form – they only increase probabilities of seeing in particular ways. Texture and other features are rejected or ignored for highly improbable objects; such as the hollow mould of a face – which looks like a normal face though it is hollow. This takes us again to the notion of betting on the odds: unlikely objects require more data – more cues – for correct perception.

Shading and shadow

Artists make remarkably effective use of many of the visual cues to depth. In a pencil drawing, shading may be used to indicate the form of an object. The shading is often conventional stippling, or equally spaced lines indicating a flat region, and unequal spacing to indicate that the surface is sloping or irregular.

Shading may also indicate shadow, and this is a different matter from regularity of surface texture. Shadows indicate the direction of

10.15 The apparatus for producing the effect opposite. The nearer square and the playing card are cut away so as not, in fact, to overlap the more distant square and card. This is so unlikely that the distant and further squares and cards are reversed in distance. This shows that overlap of further by nearer objects is an important depth cue, and it involves some knowledge of the world of objects.

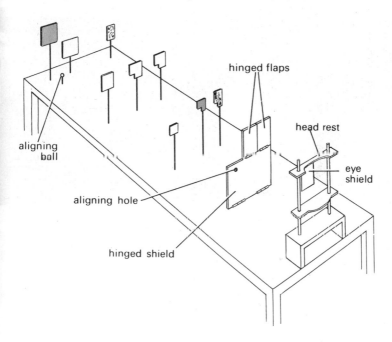

light falling on the objects, and also where a second object obstructs the light. The shadow may be cast by overhanging features of an object – as when texture is revealed by shadow – and then both the texture of the surface and the direction of the illumination are indicated by the form and direction of shadows. This is a matter of surprising importance. Shadows are important for they supplement the single eye, to give something surprisingly close to binocular depth. The light-source revealed by the shadows replaces the missing eye of the painter.

Consider a portrait taken full face, but with strong side lighting.

10.16 Two viewing positions from one camera angle —
the shadow shows the profile of the nose and eyes.

SHADOW

10.17 Letters? These are merely shadows, but we see the objects
which would cast the shadows. Look carefully: there are not
really raised letters casting the shadows, though we 'see' them.
The brain sometimes invents missing objects when they are
highly probable — even on shadow evidence.

The profile form of the nose is actually shown on the cheek (figure 10.16). The shadow thus gives us a second view of the nose. We get the same effect when looking at the moon through a telescope – indeed until space travel our knowledge of the profiles of crater walls and mountains depended on seeing their shadows cast by oblique sunlight. It is possible to measure the lengths of the shadows, and deduce accurately the heights and forms of lunar features. For object perception the visual system does this continually, and it is important – the world looks rather flat when the light is behind us for then there are no shadows.

We have already noted that perceived depth can be reversed by interchanging the eyes optically, each eye receiving the normal view of the other (see chapter 5). Interestingly enough, reversal in depth given by the light-source 'eye' casting shadows can also occur if it is shifted from its usual position. The point is that light normally falls from above: the sun cannot shine from below the horizon, and artificial light is generally placed high. When, however, illumination is from below, we tend to get reversed depth, much as when our eyes are switched by a pseudoscope (figure 5.12).

This effect was noted by several early writers: David Brewster (1781–1868) records it in his *Letters on Natural Magic*, where he describes how, when the direction of light falling upon a medal is changed from above to below, depressions become elevations and elevations depressions – i.e. intaglios become cameos and *vice versa*. This was observed at an early meeting of the Royal Society, by a member who was looking at a guinea coin through a microscope. Brewster said of it:

The illusion . . . is the result of the operation of our own minds, whereby we judge the forms of bodies by the knowledge we have acquired of light and shadow.

He then goes on to experiment with the effect, finding it more marked in adults than in children. He found that visual depth may become reversed even when true depth is indicated by touch. This must rank as one of the earliest psychological experiments. It seems that chickens are subject to the same effect; for they peck at the wrong distance with reversed lighting. For them this is innate.

The effect can occur when observing the moon through a telescope

10.18 (Above) An egg crate, illuminated from above. This is seen in marked depth. (Right) The same, illuminated from below — depth is reversed. (Try reversing the book; depth usually changes).

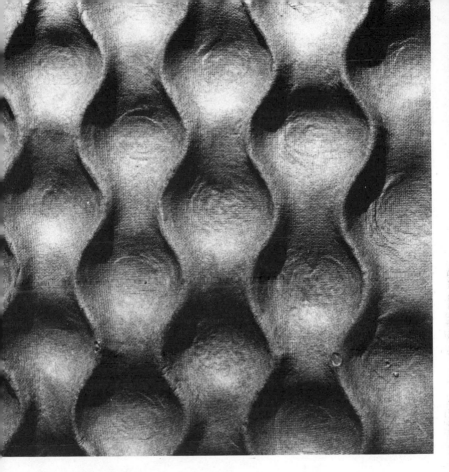

as the sun can illuminate the moon from below our horizon, though it cannot do so for objects on Earth (figure 10.18).

Although shadows are joined to and so are part of objects, they are perceptually quite distinct and very seldom confused with them. Shadows are so powerful as cues that they can evoke perception of objects even when there are no objects present. This is very clear in the typeface in figure 10.17. Here we see letters as large as life, but in fact only the shadows of imaginary letters are present. Could this effect, sometimes, make us see ghosts – figures perceptually invented to 'explain' shadows?

11.1 The 'visual cliff'. This experiment, designed by Mrs Eleanor Gibson, subjects babies or young animals to a drop under a glass sheet.
The baby refuses to crawl on the glass, over the drop, and so evidently sees the depth and the danger.

11 Do we have to learn how to see?

A most ancient question in philosophy is: How do we come to know the world? Indeed, philosophers are divided into those termed metaphysical – who hold that we have knowledge of the world apart from any sensory experience – and empiricists – who claim that all knowledge is derived from observation and measurement. To the metaphysician it seems clear that by sitting in an armchair and thinking sufficiently hard, and in the right kind of way, it is possible to make even 'contingent' discoveries, such as the number of planets, without having to look. To empiricists, perception is basic for all knowledge: – though we do find that perceptions are far from being neutral or free from errors of many kinds.

For 2,000 years metaphysicians upheld their claim by pointing to mathematics, especially geometry, where new facts were continually discovered not by experiment or by observation, but by thinking and juggling with mathematical symbols. It is only since Karl Gauss (1777–1855) that it has become clear that mathematical discoveries are of a special kind: constituting knowledge not of objects, but of the allowed arrangements of symbols. Mathematical discoveries are about mathematics, not about the world. Gauss was among the first to realise that there is not only one possible geometry: other geometries can be invented – and it is an empirical question which best fits our world. Mathematics is useful in making explicit the steps of an argument; and for making the steps between problem and conclusion automatic (given that a suitable method has been found) and in presenting data in convenient forms. But neither mathematics nor logic give new facts about the world, in the sense that facts are discovered by observation. All knowledge starts from experience.

There are, however, many animals which seem to know a lot about the world of objects before ever they experience it. Insects play successful hide-and-seek with predator and prey before they have time to learn. Migrating birds use the pattern of the stars to guide them over featureless oceans, even when they have never seen the sky.

How can these things happen if the empiricists are right that all knowledge is derived from the senses?

Experimental psychology has grown up from philosophy, and the smoke and the ashes of ancient controversy cling to it still. Psychologists distinguish between innate and learned responses; the former implying knowledge without previous experience, the latter knowledge from observation. But the issue in psychology is not quite the same as it is to the philosophers. To philosophers the question is: Can we know before we have perceived? To the psychologist the question is also: Can we perceive before we have learned how to perceive? If perception is a skill, do we have to learn this skill? How, though, could we learn to see before we have seen anything? Animals, including insects, can respond appropriately to some objects upon first encounter, but this does not make them metaphysicians. They are heirs, by inheritance, to a state of knowledge won by ancestral disasters. What is this inherited knowledge?

What is learned by an individual cannot be directly inherited by its descendants, but the genetic coding can become modified, by the processes of natural selection, to give the capacity to respond appropriately to some objects, or situations, encountered for the first time by the individual. Patterns of behaviour, and the ability to recognise objects such as long-standing enemies, are as important to the survival of a creature as is its structure. Indeed, the limbs and the senses are useless unless they are used to effect: just as useless as tools without the skill to guide and direct them. Just as the simple reflexes serve, without learning, to protect a young animal from danger of falling, or being suffocated, so there are some inborn perceptual skills for situations which were important for the survival of many generations of ancestors. This inheritance gives a basis for perception.

Animals low down on the evolutionary scale rely almost entirely upon unlearned perception of objects. But their perceptual range is small, and they respond only in stereotyped ways. Some insects do show perceptual learning; but for them the emphasis is upon 'innate' knowledge, learning being restricted largely to the whereabouts of their hive, or other 'home' – which could not be known innately. The bee does not have to learn about flowers, for these are ancient objects learned by evolutionary changes of the nervous system through many generations. The bee sucks where ancestors found nectar – for by nectar they survived. The pattern of the petals leading to nectar became built into the bee's brain, as those without it died for lack of honey.

Given that structure develops by natural selection, it is not surprising that the same is true for some behaviour and perception. What would be truly surprising in an empiricist view of nature, would be to find immediate 'recognition' of artificial or unimportant shapes. For example, if a child was found to recognise a language without having been taught it this would be startling, for the knowledge could not have become genetically coded. But there is no good evidence for this kind of innate immediate knowledge. The point here may seem obvious; but not so long ago metaphysicians did indeed seriously hold that by pure thought the number of the planets could be known without the need for observation. It was this assumption that then seemed obvious, while the empiricist's position which we accept seemed absurd and to flout the facts.

Recovery from blindness

It might seem a simple matter to bring up animals in darkness: to deny them vision for months or years and then discover what they see given light. Pioneering experiments of this kind were undertaken by R. L. Reissen. He found severe behavioural losses; but some of these might be due to degeneration of the retina (which was found to occur in darkness, though less so with diffusing goggles) and also to the remarkably passive state of animals, especially monkeys, reared in the dark. It was extremely difficult to infer specific perceptual changes or loss from the general lack of behaviour of these animals. It is not socially possible to bring up human babies in the dark – but there are cases of adult recovery from blindness. Can these cases tell us how perception develops?

The great American psychologist William James (1842–1910) described the world of the baby as 'a blooming, buzzing confusion' but is this so? How can we find out what the visual world of the baby is like? This question has engaged the attention of philosophers who have been fascinated by the possibility that one might learn how a baby comes to see by somehow giving sight to a blind man, and asking him what he sees before he has had time to learn. Although operations for recovery of sight from cataract are common such operations are rare for congenital blindness; but there are such cases as we shall see in a moment.

Perception of the blind was described by Descartes, in the *Dioptrics* (1637). He imagines a blind man discovering the world with a stick:

... without long practice this kind of sensation is rather confused and dim; but if you take men born blind, who have made use of such sensations all their life, you will find they feel things with such perfect exactness that one might almost say they see with their hands.

The implication is that this kind of learning might be necessary for the normal child to develop his world of sight.

John Locke (1632–1704) received a celebrated letter from his friend Molyneux which posed the question:

Suppose a man born blind, and now adult, and taught by his touch to distinguish between a cube and a sphere of the same metal, and nighly of the same bigness, so as to tell, when he felt one and the other, which is the cube, which is the sphere. Suppose then the cube and the sphere placed on a table, and the blind man made to see; query, Whether by his sight, before he touched them, he could now distinguish and tell which is the globe, which the cube? To which the acute and judicious proposer answers: 'Not. For though he has obtained the experience of how a globe, how a cube, affects his touch; yet he has not yet attained the experience, that what affects his touch so or so, must affect his sight so or so. . . .'

Locke comments (in *Essay Concerning Human Understanding*, 1690) as follows:

I agree with this thinking gentleman, whom I am proud to call my friend, in his answer to this problem; and am of the opinion that the blind man, at first, would not be able with certainly to say which was the globe, which the cube . . .

Here is a suggested psychological experiment: with a guessed result.

George Berkeley (1685–1753), the Irish philosopher, also considered the problem. He says:

In order to disentangle our minds from whatever prejudices we may entertain with the relation to the subject in hand nothing is more apposite than the taking into our thoughts the case of one born blind, and afterwards, when grown up, made to see. And though perhaps it may not be an easy task to divest ourselves entirely of the experience received from sight so as to be able to put our thoughts exactly in the posture of such a one's: we must nevertheless, as far as possible, endeavour to frame conceptions of what might reasonably be supposed to pass in his mind.

Berkeley goes on to say that we should expect such a man not to know that anything was:

high or low, erect or inverted . . . for the objects to which he had hitherto used to apply the terms up and down, high and low, were such only as affected or were some way perceived by his touch; but the proper objects of vision make new sets of ideas, perfectly distinct and different from the former and which can by no sort make themselves perceived by touch.

Berkeley then goes on to say that his opinion is that it would take some time to learn to associate touch with vision. This is a clear statement of the need for experience in infancy before vision is possible, which is generally stressed by empiricist philosophers. It is also a psychological statement – which could well be false – that touch and vision are neurally separate though they *seem* very different.

There have been several actual cases of the kind imagined by Molyneux. The most famous is that of a thirteen-year-old boy, described by Cheseldon in 1728. There are now nearly a hundred cases, though few are adequately described, with sufficient evidence of early lack of vision, to be useful. The first dates from AD 1020, and there are some recent cases with various new kinds of eye operation.

Some of the reported cases are much as the empiricist philosophers expected. They could see but little at first, being unable to name or distinguish between even simple objects or shapes. Sometimes there was a long period of training before they came to have useful vision, which indeed in many cases was never attained. Some gave up the attempt and reverted to a life of blindness, often after a period of severe emotional disturbance. On the other hand, some did see quite well almost immediately, particularly those who were intelligent and active, and who had received a good education while blind. The overall difficulty which these people have in naming the simplest objects by sight, and the slowness in the development of perception, so impressed the Canadian psychologist D. O. Hebb that he gave a lot of weight to this evidence, suggesting that indeed it shows how important perceptual learning is to the human infant.

It is important to note, however, that the reported cases do not all show extreme difficulty, or slowness, in coming to see. We should also remember that the operation itself is bound to disturb the optics of the eye, so that we cannot expect a reasonable image until the eye has had time to settle down after the operation. This is particularly important in the case of the removal of the lens for cataract, which constitutes all the earlier cases, while the other kind of operable blindness – opacity of the cornea – involves less change or damage, and more rapidly gives an adequate retinal image. We shall now discuss in some detail such a case, which I had the good fortune to investigate, with my colleague Jean Wallace in 1961–2.

The case of S.B. This case, a man of fifty-two, whom we may call S.B., was when blind an active and intelligent man. He would go for cycle rides, with a friend holding his shoulder to guide him; he would often dispense with the usual white stick, sometimes walking into things such as parked cars, and hurting himself. He liked making things, with simple tools in a shed in his garden. All his life he tried to picture the world of sight: he would wash his brother-in-law's car, imagining its shape as vividly as he could. He longed for the day when he might see, though his eyes had been given up as hopeless, so that no surgeon would risk wasting a donated cornea. Finally the attempt was made, (when cornea 'banks' were set up) and it was successful. But though the operation was a success, the story ends in tragedy.

When bandages were first removed from his eyes, so that he was no longer blind, he heard the voice of the surgeon. He turned to the voice, and saw nothing but a blur. He realised that this must be a face, because of the voice, but he could not see it. He did not suddenly see the world of objects as we do when we open our eyes.

But within a few days he could use his eyes to good effect. He could walk along the hospital corridors without recourse to touch; he could even tell the time from a large wall clock, having all his life carried a pocket watch with no glass, so that he could feel the time from its hands. He would get up at dawn, and watch from his window the cars and trucks pass by. He was delighted with his progress, which was extremely rapid.

When he left the hospital, we took him to London and showed him many things he never knew from touch, but he became curiously dispirited. At the zoo he was able to name most of the animals correctly, having stroked pet animals and enquired how other animals differed from the cats and dogs he knew by touch. He was also of course familiar with toys and models. He certainly used his previous knowledge from touch, and reports from sighted people to help him name objects by sight, which he did largely by seeking their characteristic features. But he found the world drab, and was upset by flaking paint and blemishes on things. He liked bright colours, but became depressed when the light faded. His depressions became marked, and general. He gradually gave up active living, and three years later he died.

Depression in people recovering sight after many years of blindness seems to be a common feature of the cases. Its cause is probably complex, but in part it seems to be a realisation of what they have missed – not only visual experience, but opportunities to do things denied them during the years of blindness. Some of these people revert very soon to living without light, making no attempt to see. S.B. would often not trouble to turn on the light in the evening, but would sit in darkness.

We tried to discover what his visual world was like, by asking him questions and giving him various simple perceptual tests. While still in the hospital, before he became depressed, he was most careful with his judgments and his answers. We found that his perception of distance was peculiar, and this is true of earlier cases. He thought he would just be able to touch the ground below his window with his feet, if he hung from the sill with his hands; but in fact the distance down was at least ten times his height. On the other hand, he could judge distances and sizes quite accurately provided he already knew the objects by touch. Although his perception was demonstrably peculiar, he seldom expressed surprise at anything he saw. He drew the elephant (figure 11.2) before we showed him one at the zoo; but upon seeing it, he said immediately: 'there's an elephant', and said it looked much as he expected it would. On one object he did show real surprise, and this was an object he could not have known by touch – the moon. A few days after the operation, he saw what he took to be a reflection in a window (he was for the rest of his life fascinated by reflections in mirrors and would spend hours sitting before a mirror in his local public house, watching people) but this time what he saw

11.2 S.B.'s drawing of an elephant. He drew this before having seen one. Half an hour later we showed him a real elephant, at the London zoo, and he was not at all surprised by it!

was not a reflection, but the quarter moon. He asked the Matron what it was, and when she told him, he said he had thought the quarter moon would look like a quarter piece of cake!

S.B. never learned to read by sight (he read Braille, having been taught it at the blind school) but we found that he could recognise block capital letters, and numbers, by sight without any special training. This surprised us greatly. It turned out that he had been taught upper case, though not lower case, letters at the blind school. They were given raised letters on wooden blocks, which were learned by touch. Although he read upper case block letters immediately by sight, it took him a long time to learn lower case letters, and he never managed to read more than simple words. Now this finding that he could immediately read letters visually which he had already learned by touch, showed very clearly that he was able to use his previous touch experience for his new-found vision. This is interesting to the psychologist, for it indicates that the brain is not so departmentalised as is sometimes thought. But it makes any finding of these cases difficult or impossible to apply to the normal case of a human infant coming to see. The blind adult knows a great deal about the world of objects through touch and hearsay: he can use some of this information to help him identify objects from the slightest cues. He also has to come to accept and trust his new sense, which means giving up the habits of many years. His case is really quite unlike that of the child's.

S.B.'s use of early touch experience comes out clearly in drawings which he did for us, starting while still in the hospital and continuing for a year or more. The series of drawings of buses (in figure 11.3) illustrate how he was unable to draw anything he did not already know by touch. In the first drawing the wheels have spokes, and spokes are a distinctive touch feature of wheels. The windows seem to be represented as he knew them by touch, from the inside. Most striking is the complete absence of the front of the bus, which he would not have been able to explore with his hands, and which he was still unable to draw six months or even a year later. The gradual introduction of writing in the drawings indicates visual learning: the sophisticated script of the last drawing meant nothing to him for nearly a year after the operation, although he could recognise block capitals while still in the hospital, having learned them previously from touch. It seems that S.B. made immediate use of his earlier touch experience, and that for a long time his vision was limited to what he already knew.

We saw in a dramatic form the difficulty that S.B. had in trusting and coming to use his vision, whenever he had to cross the road. Before the operations he was undaunted by traffic. He would cross alone, holding his arm or his stick stubbornly before him, when the traffic would subside as the waters before Christ. But after the operation, it took two of us on either side to force him across a road: he was terrified as never before in his life.

When he was just out of the hospital, and his depression was but occasional, he would sometimes prefer to use touch alone, when identifying objects. We showed him a simple lathe (a tool he had wished he could use) and he was very excited. We showed it him first in a glass case, at the Science Museum in London, and then we opened the case. With the case closed, he was quite unable to say anything about it, except that the nearest part might be a handle (which it was – the transverse feed handle), but when he was allowed

11.3 (Below) S.B.'s first drawing of a bus (48 days after the corneal graft operation giving him sight). All the features given were probably known to him by touch. The front, which he had not explored by touch, is missing, and he could not add it when we asked him to try.
(Top right) Six months later. Now he adds writing, the 'touch' spokes of the wheels have been rejected, but he still cannot draw the front.
(Bottom right) A year later the front is still missing. The writing is sophisticated, though he could hardly read.

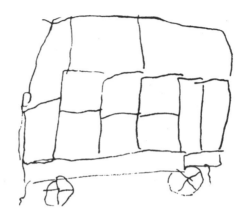

to touch it, he closed his eyes and placed his hand on it, when he immediately said with assurance that it was a handle. He ran his hands eagerly over the rest of the lathe, with his eyes tight shut for a minute or so; then he stood back a little, and opening his eyes and staring at it he said: 'Now that I've felt it I can see.'

Although many philosophers and psychologists think that these cases can tell us about normal perceptual development in infants, I am inclined to think that they tell us rather little. As we have seen, the difficulty is essentially that the adult, with his great store of knowledge from the other senses, and reports from sighted people, is very different from the infant who starts with no knowledge from experience. It is extremely difficult, if not entirely impossible, to use these cases to answer Molyneux's question. The cases are interesting and dramatic, but when all is said, they tell us little about the world of the baby, for adults with restored vision are not living fossils of infants.

Recently sight has been restored by implanting acrylic lenses.

198

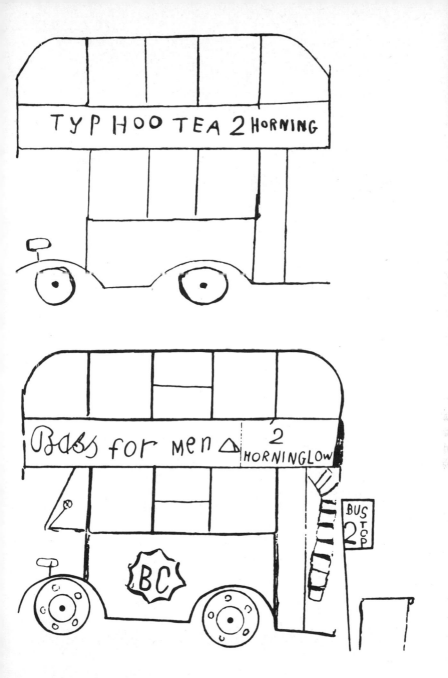

11.4 Fantz's apparatus (right) for observing babies' eye movements while they are shown various designs, or objects. Here the baby is shown an illuminated ball, while the eye positions are photographed.
(Below) A simple face, and a randomised face-like design which were shown to very young babies. They spent longer looking at the true face picture (as judged by their eye movements).

Rejection by the eye's tissue is avoided, by placing the lens in a tooth extracted from the patient; the tooth being implanted in the eye, with the lens placed in a hole drilled through the tooth. Many of the S.B. findings have been confirmed with these new cases. This extraordinary operation, from which people walk around with a tooth in their eyes, as in a Greek myth, is performed in Italy. The results have been reported by Alberto Valvo (1971).

Direct evidence from babies

To get direct evidence from babies is difficult because they have little motor co-ordination and no language. They do however have co-ordinated eye movements within a few weeks of birth. R. L. Fantz discovered that babies spend about twice as long fixating a face-like

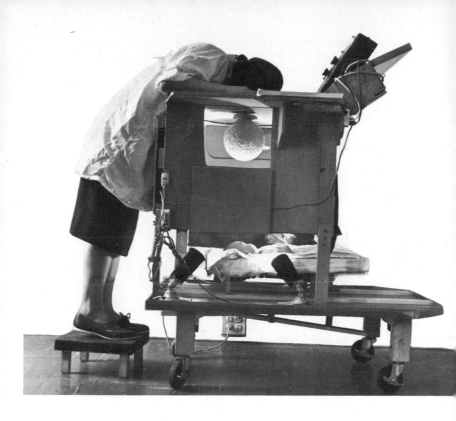

picture than a 'jumbled' picture of the same features (figure 11.4 and figure 11.5). This may imply innate recognition of faces; but very early learning is not ruled out as the mother's face has not been hidden from the baby.

It is also found from their eye movements that babies seem to prefer solid objects to flat representations of the same objects, so possibly they have some innate appreciation of depth.

Mrs Eleanor Gibson, while picnicking on the rim of the Grand Canyon, wondered whether a young baby would fall off. This thought led her to a most elegant experiment, for which she devised a miniature and safe Grand Canyon. The apparatus is shown in figure 11.1, which shows a central 'bridge' with, on one side a normal solid floor, and over the drop a large sheet of strong glass. An infant (or in other experiments a young animal) is placed on the central bridge.

11.5 Some results of Fantz's eye movement experiments on babies. The horizontal bars show the relative times they spent looking at the various designs shown on the left of the diagram.

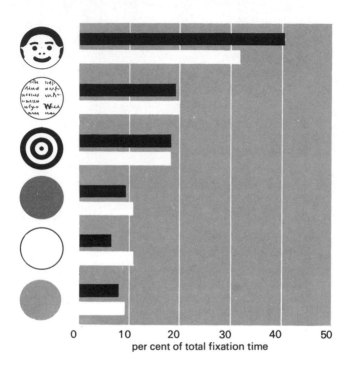

The question is: will he crawl on the glass over the drop? The answer is that the baby will not leave the bridge on the drop side, and cannot be enticed over it by his mother shaking his rattle, though he will crawl quite happily on to the normal floor on the other side of the bridge. It thus seems that babies at the crawling stage can appreciate a drop, though probably only from motion parallax.

Recent experiments of great interest by J. S. Bruner and several others show that babies a few weeks old have surprising perceptual abilities. By making video taped records of babies reaching for toys or small wooden bricks, Jerome Bruner has found that they open their fingers appropriately to the size of the object. This implies that they can use vision to judge size, and also direction and distance, of nearby objects. There is also evidence suggesting that babies have some expectations about objects from visual information. By projecting images of objects stereoscopically, so that they appear to lie a foot or so away though there is nothing to touch, T. C. Bower found that babies seem to show surprise and be upset when they reach out to grasp these images having no substance. It is difficult to be sure that the babies are surprised; but if so, they have expectations of touch from visual information, and they have some knowledge of non-visual properties of objects, such as distance and size. There is no evidence for innate knowledge of *particular* kinds of objects except faces. The breast is probably recognised by smell. Babies are easily distracted from visually interesting objects by new or significant scents, which adds another difficulty to these experiments.

Development of 'object persistence'   This takes us to important questions, such as: What do babies know and understand? By what stages do they gain understanding of things around them? How much do they learn for themselves, and how much do they need adult help? As we have stressed throughout, perception is far more than responding to stimuli. Perception is much more a matter of acting appropriately to *sources* of stimuli. In short, we should consider not merely *pattern* recognition but *object* recognition – objects having lawful pasts and futures, and important unsensed characteristics. The baby's task is to discover and know about objects from his limited sensory data. The experiments most directly concerned with this are on 'object persistence'.

If a baby follows an object of interest with his eyes, and the object passes behind an opaque screen, his eyes tend to move in anticipation of the object emerging. If it does not emerge, he may appear surprised – so there is evidence that the baby has expectations of the permanence of the object though the visual stimulus is lost. If the object is replaced by another object while behind the screen – so that a teddy bear disappears behind the screen and a toy fire engine emerges – very young babies show no surprise; but when a year or so of age they show consternation. This indicates development of knowledge of particular kinds of objects with experience.

In the few studies where human babies have been brought up with baby apes, treated as nearly as possible the same, in a human family, the ape initially develops faster; but is overtaken by the human baby at about the age the human baby starts to talk. It is then that the human develops abilities denied for ever to the ape and to all other animals. Is there a connection between human language and human perceptual ability? Is language possible for humans because of our sophisticated knowledge-based perception? Or, is language a tool for classifying knowledge in subtle ways to increase the power of perception? Our classifications by words do seem to parallel classes of objects as we see them, and providing names for new kinds of objects (such as microscopic structures) can probably aid new perceptual learning in adults. We do not know how, or if at all, language is related to perception; but the notion of Noam Chomsky, that all natural human languages have a common Deep Structure (though this is not obvious from the great differences between 'surface' grammars) is suggestive. Reading is itself a special kind of perception, which we do not at all well understand, but like other perception it is both knowledge-based and adds to our knowledge given innately and by our experience of the world.

Adapting to displaced images

Although experiments with babies are progressing well, their behaviour is so limited that we cannot say much more, as yet. To pursue perceptual learning further we must look at experiments on adults adapting to peculiar situations. We will start with the classical work of G. M. Stratton. He wore inverting lenses for days on end – and was the first man to have retinal images not upside down!

Stratton devised a variety of optical devices for displacing and inverting the retinal image. He used lens and mirror systems, including special telescopes mounted on spectacle frames so that they could be worn continuously. These lenses inverted both vertically and horizontally. Stratton found that if a pair of inverting lenses was worn giving binocular vision the strain was too great as normal convergence was upset. He therefore wore a reversing telescope on but one eye, keeping the other eye covered. When not wearing the inverted lenses he would keep both eyes covered. At first objects seemed illusory and unreal. Stratton wrote:

... the memory images brought over from normal vision still continued to be the standard and criterion of reality. Things were thus seen in one way and thought of in a far different way. This held true also for my body. For the parts of my body were felt to be where they would have appeared had the instrument (the inverting lenses) been removed; they were seen to be in another position. But the older tactual and visual localisation was still the real localisation.

Later, however, objects would look almost normal.

Stratton's first experiment lasted three days, during which time he wore the 'instrument' for about twenty-one hours. He concluded:

I might almost say that the main problem – that of the importance of the inversion of the retinal image for upright vision – had received from the experiment a full solution. For if the inversion of the retinal image were absolutely necessary for upright vision . . . it is difficult to understand how the scene as a whole could even temporarily have appeared upright when the retinal image was not inverted.

Objects only occasionally looked normal, however, and so Stratton undertook a second experiment with his monocular inverting arrangement, this time wearing it for eight days. On the *third day* he wrote:

Walking through the narrow spaces between pieces of furniture required much less care than hitherto. I could watch my hands as they wrote, without hesitating or becoming embarrassed thereby.

On the *fourth day* he found it easier to select the correct hand, which had proved particularly difficult.

When I looked at my legs and arms, or even when I reinforced my effort of attention on the new visual representation, then what I saw seemed rather upright than inverted.

By the *fifth day* Stratton could walk around the house with ease. When he was moving around actively, things seemed almost normal, but when he gave them careful examination they tended to be inverted. Parts of his own body seemed in the wrong place, particularly his shoulders, which of course he could not see. But by the evening of the *seventh day* he enjoyed for the first time the beauty of the scene on his evening walk.

On the *eighth day* he removed the inverting spectacles, and found that

... the scene had a strange familiarity. The visual arrangement was immediately recognised as the old one of pre-experimental days; yet the reversal of everything from the order to which I had grown accustomed during the last week, gave the scene a surprising bewildering air which lasted for several hours. It was hardly the feeling, though, that things were upside down.

One has the impression when reading the accounts of Stratton, and the investigators who followed him, that there is always something queer about their visual world, though they have the greatest difficulty in saying just what is wrong with it. Perhaps rather than their inverted world becoming normal, they cease to notice how odd it is, until their attention is drawn to some special feature, when it does look clearly wrong. We read of such situations where writing appears in the right place in the visual field and at first sight looks like normal writing, except that when one attempts to read, it is seen as inverted.

Stratton went on to perform other experiments which though less well-known are just as interesting. He devised a mirror arrangement which, mounted in a harness (figure 11.6), visually displaced his own body, so that it appeared horizontally in front of him, and at the height of his own eyes. Stratton wore this mirror arrangement for three days (about twenty-four hours of vision) and he reported:

I had the feeling that I was mentally outside my own body. It was, of course, but a passing impression, but it came several times and was vivid while it lasted ... But the moment critical interest arose, the simplicity of the state was gone, and my visible actions were accompanied by a kind of wraith of themselves in the older visual terms.

Stratton summed up his work in the following words:

The different sense-perceptions, whatever may be the ultimate course of their extension, are organised into one harmonious spatial system. The harmony is found to consist in having our experiences meet our expectations ... The essential conditions of the harmony are merely those which are necessary to build up a reliable cross-reference between the two senses. This view, which was first based on the results with the inverting senses, is now given wider interpretation, since it seems evident from the later experiment that a given tactual position may have its correlated visual place not only in any direction, but also at any distance in the visual field.

11.6 Stratton's experiment, in which he saw himself suspended in space before his eyes, in a mirror. He went for country walks wearing this arrangement.

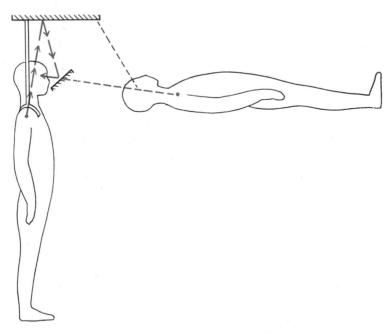

Several investigators have followed up Stratton's work. G. C. Brown used prisms to rotate the field of both eyes through 75°, and found that while this reduced the efficiency of depth perception, there was little or no evidence that it improved with experience, though he and his subjects did find that they got used to their tilted world. Later studies are those of P. H. Ewert, who repeated Stratton's experiment, using a pair of inverting lenses in spite of the strain on the eyes found by Stratton. Ewert's work has the great merit that he made systematic and objective measures of his subject's ability to locate objects. He concluded that Stratton somewhat exaggerated the amount of adaptation that occurred, and this led to a controversy that is still unresolved.

207

11.7 Ivo Kohler finds that after directing the eyes through
the green filter to one side and the red filter to the other
they become adapted and compensate to the filters for
the viewing positions. When the filters are removed
the side which was green looks red and vice versa.
This adaptation must be in the brain and not in the eyes.

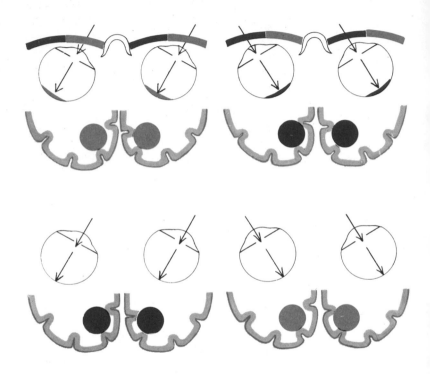

The problem was taken up by J. and J. K. Paterson, using a
binocular system similar to Ewert's. After fourteen days they did not
find complete adaptation to the situation. Upon re-testing the subject
of the experiment eight months later, they found that when the lenses
were worn, the subject immediately showed the various modifications
to his behaviour which he previously developed wearing the inverting
spectacles. It thus seemed that the learning consisted of a series of
specific adaptations overlying the original perception, rather than a
reorganisation of the entire perceptual system.

The most extensive recent experiments on humans have been carried out by T. Erisman followed by Ivo Kohler. Kohler and his subjects wore their reversing spectacles for long periods. Both Stratton's and Kohler's experiments rely on verbal reports. Kohler stresses the 'inner world' of perception, following the European tradition which we find in the German Gestalt writers, and in the more recent work of Michotte on the perception of causality (chapter 12). This emphasis is foreign to the Behaviourist tradition of America, and certainly it is unfortunate that but little precise recording of the subject's movements during the experiments was attempted. From the verbal reports it is difficult to imagine the 'adapted' world of the experimental subjects, for their perceptions seem to be curiously shuffled and even paradoxical. For example, pedestrians were sometimes seen on the correct side of the street, when the images were right–left reversed, though their clothes were seen as *the wrong way round*! Writing is one of the more puzzling things observed. With right–left reversal, a scene would come to look correct except that writing remained 'mirror-writing' – right–left reversed and difficult to read.

Touch had important effects on vision: during the early stages of adaptation, objects would tend to look suddenly normal when touched, and they would also tend to look normal when the reversed appearance was physically impossible. For example a candle would look upside down until lighted, when it would suddenly look normal, the flame going upwards.

These experiments gave place to several studies of animals fitted with goggles of various kinds. Inverting goggles placed on a monkey had the effect of immobilising her for several days, so that she simply refused to move. When she did finally move it was backwards – a point of some interest as these inverting goggles tend to reverse depth perception. Similar experiments have also been tried in chickens and hens. Right–left reversing prisms were attached to the eyes of hens by M. H. Pfister, who then observed their ability to peck grain. The hens were severely disturbed, and they showed no real improvement after three months wearing the prisms. This same lack of adaptation has also been found in amphibia investigated by R. W. Sperry. With their eyes rotated through 180°, it was found that they would move their tongue in the wrong direction for food, and they would have

209

starved to death had they been left to fend for themselves. Similar results were also obtained by A. Hess, with chickens wearing wedge prisms which did not reverse the images, but shifted them by 7° either to the right or to the left. Hess found that these chickens would always peck to the side of grain, and that they never adapted to the shift of the image caused by the wedge prisms (figure 11.8). Hess concludes from these experiments:

Apparently the innate picture which the chick has of the location of objects in its visual world cannot be modified through learning if what is required is that the chick learns to perform a response which is antagonistic to its instinctive one.

It seems quite clear from the various experiments, that animals show far less adaptation to shift or reversal of the image than do

11.9 Apparatus designed by Held and Hein to discover whether perceptual learning takes place in a passive animal. The kitten on the right is carried about by the active kitten on the left. They thus have similar visual stimulation. Following visual experience limited to this situation, only the active animal is able to perform visual tasks – the passive animal remains effectively blind.

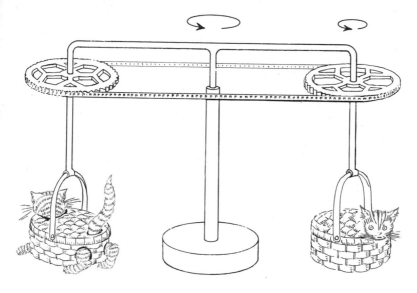

human observers. Indeed, only monkeys and men show any perceptual adaptation to these drastic changes.

There is recent evidence, mainly from the work of R. Held and his associates, particularly A. Hein, to show that for compensation to displaced images to occur it is essential that the subject should make active corrective movements. Held considers that active movement is vital for compensations, and also that it is essential for such perceptual learning in the first place. One of their experiments with kittens is particularly ingenious and interesting. They brought up pairs of kittens in darkness; with vision only in the experimental situation, in which one kitten served as a control for the other. The two kittens were placed in baskets attached to opposite ends of a pivoted beam, which could swing round its centre, while the baskets could also rotate. It was so arranged that a rotation of one basket caused the other to rotate similarly (the arrangement is seen in figure 11.9). With this ingenious device, both kittens received much the same visual stimulation. One of the kittens was carried *passively* in its basket; the other kitten, whose limbs could touch the floor, moved the apparatus around *actively*. Thus one kitten was passive and the other active though both had similar visual stimulation. Held found that only the active kitten gave evidence of perception, the passive animal remaining effectively blind. But is this 'blindness' absence of correlations built up between its vision and its behaviour? Could it indeed be seeing – but be unable to let us know that it sees?

Richard Held has also undertaken human experiments with deviating prisms, finding that active arm movement (striking a target with the finger) is necessary for effective adaptation. Is this adaptation *perceptual* or is it *proprioceptive* – in the control system of the limbs? The principal supporter of proprioceptive adaptation is C. S. Harris. This cannot however apply to adaptation to *distortions* of vision, which we shall go on to discuss.

So far we have considered experiments on inverting and tilting the images, but other kinds of disturbance can be produced. These are important because they involve some internal reorganisation in the perceptual system itself, rather than simple changes in the relation between the worlds of touch and vision. This can be done by wearing special lenses which *distort*, rather than displace, the image on the retina.

J. J. Gibson found, while undertaking an experiment wearing prisms to deviate the field to one side (15° to the right), that the *distortion* of the image, which such prisms inevitably produce in addition to the shift, gradually became less marked while he wore the prisms. He went on to make accurate measures of the adaptation to the curvature produced by the prisms, and he found that the effect diminished although his eyes moved about freely. In fact, the adaptation was slightly more marked with free inspection of the figure with eye movements, than when the eyes were held as still as possible.

There is another kind of adaptation, at first sight similar to that found by Gibson with his distorting prisms but almost certainly different in its origin and its significance to the theory of perception. These effects are known as *figural after-effects* and they have received a great deal of experimental attention, though they remain somewhat mysterious.

Figural after-effects are induced when a figure is looked at for some time (say half a minute) with the eyes held very still. If a curved line is fixated in this way, a straight line viewed immediately afterwards will for a few seconds, appear curved in the opposite direction. The effect is similar to Gibson's, but for figural after-effects it is essential that the eyes should be still.

There are limitations to the kinds of inversion possible by simple optical means. A technique has been developed by K. U. and W. M. Smith using a television camera and monitor, arranged so that the subject sees his own hand displaced in space or time.

It is a simple matter to give either right–left or up–down reversal of the image, and eye and hand movements are not affected. In this arrangement the hand is placed to the side of the subject behind a curtain, so that it cannot be seen directly. (Since the apparatus is far from portable, the studies are limited to short experimental sessions, rather than to continuous reversal of many days' duration.) In addition to reversals, the camera may be placed in any position, giving a view displaced in space. Using various lenses and camera distances the size may be varied, and distortions may be introduced (figure 11.10).

These techniques show that pure up–down reversal generally proves more disturbing than left–right reversal; though combined up–down and left–right is less disturbing than either alone. Changes in size had practically no effect on ability to draw objects, or on handwriting.

A remarkable discovery was made by Ivo Kohler. He wore glasses which did not distort, but which were coloured half red and half green, so that everything looked red while looking to the left and green when looking to the right (figure 11.7). Kohler found that the colours gradually weakened, and when the glasses were removed, *things seen with the eyes directed to the right looked red, and to the left, green.* This effect is quite different from normal after-images, for this effect is not related to position of the image on the retina, but to the position of the eye in the head, so it must be due to compensation taking place in the brain.

This might be related to a new effect discovered by Celeste

11.10 K.U. and W.M. Smith's experimental arrangement using a television camera and monitor to vary the viewing position or size of the subject's own hands. He can draw or write very well with large changes of view.

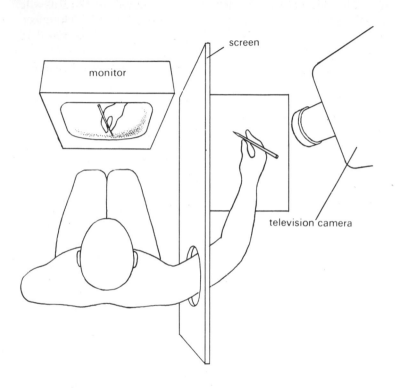

McCollough, that when lines of one orientation are repeatedly shown say red, then another orientation as say green, black and white lines of the orientations then appear coloured: as the complementaries of the 'adapting' stimuli. Similar effects occur with movements, and to many other sets of stimuli. Known as 'contingent after-effects', it is an open question whether they are due to cortical cells having shared functions, and being teased out by repeated stimulation; or whether they are due to some kind of adaptive learning during which orientation and colour become associated (see pages 110–11).

Displacement of images in time An elaboration of the television technique mentioned above makes it possible to displace retinal images not only in space, but in time. Temporal delay of images is a new kind of displacement, and promises to be of the greatest importance. The method is to use a TV camera and monitor, and to introduce a video tape recorder between the camera and monitor, with an endless tape loop so that there is a time-delay between the recording from the camera and the play-back to the monitor. The subject thus sees his hands (or any other object) in the past; the delay being set by the gap between the Record and Play-back heads (figure 11.11) for time-displaced images.

This situation is not only of theoretical interest, but is also of practical importance because controls used in flying aircraft, and operating many kinds of machine, have a delay in their action: if such delay upsets the control skill, this could be a serious matter. It was found that a short delay (about 0.5 seconds) made movements jerky and ill co-ordinated, so that drawing became almost impossible and writing quite difficult (figure 11.12). Practice gives little or no improvement.

What can we conclude?

We have had a look at a wide variety of experiments on displacements, distortions and time shifts of retinal images, as well as 'contingent after-effects' to repeated pairs of stimuli such as orientation of lines and colour which become associated. This association – unlike association by conditioning – is negative. The colours become complementary; curved lines become straighter; optical reversal is (partly and in subtle ways) cancelled. These all seem to be ways of maintaining the Constancy of perception through systematic but irrelevant change.

Is this evidence that babies have to learn how to see? It does not prove it; but evidently we do have Perceptual Learning of many various kinds even when adult. In animal experiments and some human experiments it is difficult to know whether the adaptation is perceptual or in the limb control mechanisms.

This takes us to a curious and baffling logical difficulty, which makes us question just what we should mean by 'perception',

11.11 Smith's experiment, introducing time delay between acting and seeing. The delay is given by the tape loop of the video recorder.

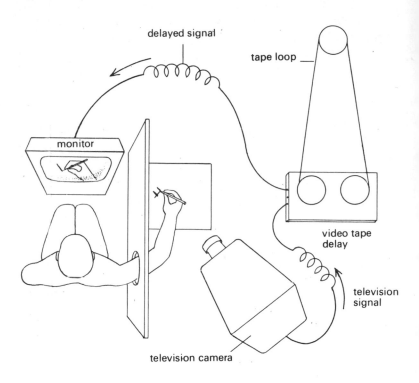

especially when interpreting animal experiments. To take for example Held and Hein's experiment with the active and passive kittens: suppose for a moment that the passive kitten *does* learn to see, in the sense that its patterns of retinal stimulation do become organised into separate objects. Now how could we known that the kitten had, in this sense, learned to see? How could he be expected to make appropriate responses if his behaviour had never become connected with his perception of objects? This raises a basic problem: should we think of perception as we know it in ourselves – *experience* of the

11.12 Drawing and writing with time delay. Left to right: normal, with TV but no delay, with TV delay. The delay provides an insuperable handicap, though displacement in space can be compensated. (The result is of practical importance, since many control tasks, such as flying, do impose a delay between action and result.)

world of objects – or should we limit its study to *behaviour* under the control of sensory information? To the strict Behaviourist, experience cannot be the subject-matter of perceptual studies: but we do all assume that in a concert hall, or a picture gallery, people experience an inner world, sufficiently important to draw them there. Whatever it is that art critics discuss, it is not only overt behaviour, but also what they experience. And yet we know nothing of the perceptual experience of animals – or whether they have experience. Whatever their behaviour, it seems insufficient as evidence of experience – consciousness – in animals or young babies. Is the problem simply lack of language? If so – is it that we require language for evidence of consciousness or is language necessary for consciousness?

Philosophers have questioned 'Other Minds' – consciousness, awareness, or sensation in other people. Since we cannot enter another's mind, how can we verify that he has a mind? How can we be certain that he has consciousness? Perhaps the best answer has been given by the great Austrian philosopher (who spent his working life at Cambridge) Ludwig Wittgenstein (1889–1951). Wittgenstein's reply to such extreme doubts as to deny mind in other people is that our language, and our experimental techniques, are not sufficiently powerful to express or to support such doubts. We cannot prove consciousness in others, but neither can we produce effective doubt to challenge what we all believe to be true. So it is rational to accept that other people have 'inner worlds' of experience.

12 Seeing and believing

The sense organs receive patterns of energy, but we seldom see merely patterns: we see objects. A pattern is a relatively meaningless arrangement of marks, but objects have a host of characteristics beyond their sensory features. They have pasts and futures; they change and influence each other, and have hidden aspects which emerge under different conditions.

A brick and a block of gelignite may look and feel much alike, but they will behave very differently. We do not generally define objects by how they appear, but rather by their uses and their causal characteristics. A table may be any one of many shapes: it is an object on which other objects may be placed and may be square or round or kidney-shaped and still be a table. For a perception to correspond to an object – to be a 'true' perception – certain expectations must be fulfilled. If a book were placed on a supposed table which then melts away, or hoots like an elephant, we would have to say that it was not, after all, a table. And also that there was not, after all, a perception – but a dream or an hallucination. The importance of regular, lawful relations in perceptions was studied by Albert Michotte at Louvain, who for many years investigated the perception of causality.

Michotte set out to discover the conditions necessary for seeing causality; by moving simple patches of colour with various velocities and time-delays, with the apparatus shown in figure 12.1. He arranged for one coloured patch to move towards and touch another, which then moves off, generally after a small controlled delay. With some combinations of velocity and delay, there is an irresistible impression that the first patch has struck the second, and pushed it, as though they were objects, such as billiard-balls.

Indeed, one experiences just this effect in a cartoon film, and the objects in a cartoon can be abstract and still display the causal relations of real objects. Michotte was inclined to think that the seeing of cause is innately given, but this view seems to be based only on the similarity of his observers' reports. We all encounter as children the

219

12.1 Michotte's apparatus for investigating perception of cause. The rotating disc carries lines. A small section of each line is viewed through a fixed slot. The visible sections move along the slot, depending on their shape on the disc. (Thus a co-axial circle would be stationary, while any other form will move.) Michotte finds that when one moves and touches another, which then moves off, its movement seems to be *caused* by the blow of the first.

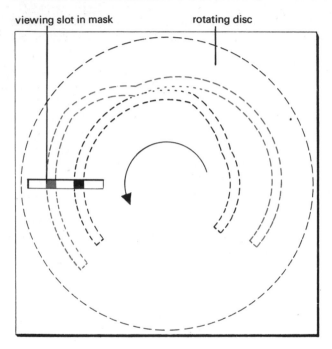

viewing slot in mask   rotating disc

same kinds of objects: so we might expect the velocity and delay characteristics of common objects to set similar perceptual appreciation of cause: so such agreement can hardly demonstrate that perception of cause is innate rather than derived from experience of objects. Experiments on observers with long experience of unusual kinds of objects might give evidence of the importance of learning for seeing cause.

Although the sensory worlds of sight, touch and smell are very different from each other, we have no hesitation in accepting that they are alternative indications of the same world of objects. But our knowledge of the world of objects is certainly not limited to sensory experience: we know about magnetism although we cannot sense it, and about atoms although they are invisible.

It seems that the retina of a frog is capable of signalling only a few characteristics; mainly movement and corners and black dots. It responds well to certain objects which are important to its survival, particularly flies, but surely its visual world must be far less rich than ours. Is our experience – and our knowledge – restricted by the limitations of our eyes and brains?

There are fish which can detect weak electric fields and local objects which distort their self-made fields. These fish have a sense entirely foreign to us, and yet we know a great deal more about electrical fields than they do; and we have learned to develop instruments which locate objects in the same way and more efficiently. Our brain has largely overcome this limitation of our sensory apparatus. Similarly, we have learned a great deal about the stars and their composition from the most meagre sensory evidence, by making deductions and using the slender evidence to test guesses and hypotheses. our eyes are general-purpose instruments for feeding the brain with comparatively undoctored information. The eyes of creatures with much simpler brains often have more elaborate eyes than ours, as they select only information essential to their survival and usable by the simple brain. It is however the freedom to make new inferences from sensory data which allows us to discover and see so much more than other animals. The large brains of mammals, and particularly humans, allow past experience and anticipation of the future to play a large part in augmenting sensory information, so that we do not perceive the world merely from the sensory information available at any given time, but rather we use this information to test hypotheses of what lies before us. Perception becomes a matter of suggesting and testing hypotheses. We surely see this process of hypothesis selecting and testing most dramatically in the *ambiguous figures*, such as the Necker cube (figure 1.4). Here the sensory input remains constant (its image may even be stabilised on the retina) and yet perception changes spontaneously, from moment to moment, as various orientations and various likely kinds of objects are selected as hypotheses of reality. Each hypothesis – each perception – is entertained alone; but none is allowed to stay when no one is better than its rivals.

Perception is however seldom ambiguous: we can live by trusting it. Fortunately these are not typical objects. Ambiguous figures put our perceptual system at a curious disadvantage; because they give no clue of which bet to make, and so it never settles for one bet. The great advantage of an active system of this kind is that it can often function in the absence of inadequate information by postulating alternative realities. But sometimes it makes wrong decisions which may be disastrous.

Figures can be drawn which apparently represent impossible objects. Several examples have been designed by L. S. and R. Penrose (figure 12.2). Simple objects can be made which appear impossible from certain viewing positions. But objects – reality – cannot be paradoxical and cannot be ambiguous. These are not properties of reality – they are due to uncertainties and conflicts in our perceptual hypotheses.

Extreme emotional stress may upset perception, much as stress can distort intellectual judgment, giving the kind of terrible but false reality expressed most dramatically by Macbeth:

> Is this a dagger which I see before me,
> The handle toward my hand? Come, let me clutch thee:
> I have thee not, and yet I see thee still.
> Art thou not, fatal vision, sensible
> To feeling as to sight? or art thou but
> A dagger of the mind, a false creation.
> Proceeding from the heat-oppressed brain?
> Mine eyes are made the fools o' the other senses,
> Or else worth all the rest: I see thee still
> And on the blade and dudgeon gouts of blood,
> Which was not so before. There's no such thing:
> It is the bloody business which informs
> Thus to mine eyes.

Why should the perceptual system be so active in seeking alternative solutions, as we see it to be in ambiguous situations? Indeed it seems more active, and more intellectually honest in refusing to stick with one of many possible solutions, than is the cerebral cortex as a whole – if we may judge by the tenacity of irrational belief in politics or religion. The perceptual system has been of biological significance for far longer than the calculating intellect.

12.2 'Impossible figures' by L. S. and R. Penrose. We see these as more than mere patterns. We are, however, incapable of seeing them as possible objects, although (as has recently been shown) objects can be made to give these projections as retinal images: when the real objects look just as impossible as the figures. (See the author's *The Intelligent Eye*, 1970.)

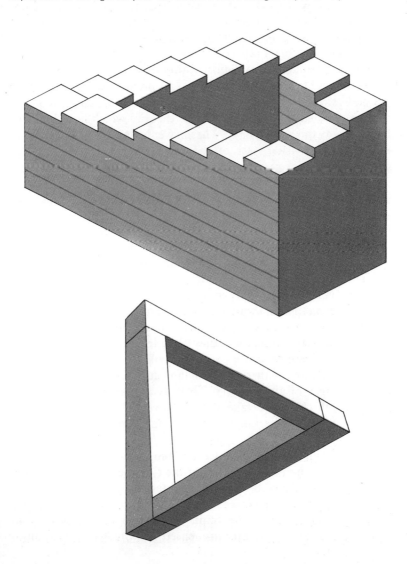

The regions of the cerebral cortex concerned with thought are comparatively juvenile. They are self-opinionated by comparison with the ancient regions of the brain giving survival by seeing.

The perceptual system does not always agree with the rational thinking cortex. For the cortex educated by physics, the moon's distance is 390,000 km. (240,000 miles); to the visual brain it is a few hundred metres. Though here the intellectual cortical view is the correct one, the visual brain is never informed, and we continue to see the moon as though it lies almost within our grasp.

The visual brain has its own logic and preferences, which are not yet understood by us cortically. Some objects look beautiful, others ugly; but we have no idea for all the theories which have been put forward why this should be so. The answer lies a long way back in the history of the visual part of the brain, and is lost to the new mechanisms which give our intellectual view of the world.

We think of perception as an active process of using information to suggest and test hypotheses. Clearly this involves learning, and whatever the final answer on the importance of perceptual learning in babies, it does seem clear that knowledge of non-visual characteristics affects how objects are seen. This is true even of people's faces: a friend or lover looks quite different from other people; a smile is not just a baring of teeth, but an invitation to share a joke. The blind man S.B. (chapter 11) never learned to interpret facial expressions; they meant nothing to him, though he could read a mood from the sound of a voice. Hunters can recognise birds in flight at incredible distance by the way each species flies: they have learned to use such subtle differences to distinguish objects which look the same to other people. We find the same with doctors diagnosing X-rays or microscope slides for signs of abnormality. There is no doubt that perceptual learning of this kind does take place; but in spite of all the evidence we still do not know for certain how far learning is required for the basis of perception.

It is not difficult to guess why the visual system has developed the ability to use non-visual information and to go beyond the immediate evidence of the senses. By building and testing hypotheses, action is directed not only to what is sensed but to what is likely to happen, and it is this that matters. The brain is in large part a probability computer; and our actions are based on predictions to future situations. Perhaps inevitably, we cannot predict with certainty what our, or other people's, predictions will be – what they will see or how they will behave. This is a price we have to pay; but intelligent behaviour is not possible without prediction. Predicting from hypotheses derived from meagre data is the hallmark of perception, intelligence and science. Science is shared perception.

If the brain were unable to fill in gaps and bet on meagre evidence, activity as a whole would come to a halt in the absence of sensory inputs. In fact we may slow down and act with caution in the dark, or in unfamiliar surroundings, but life goes on and we are not powerless to act. Of course we are more likely to make mistakes (and to suffer hallucinations or illusions) but this is a small price to pay for gaining freedom from immediate stimuli determining behaviour, as in simple animals which are helpless in unfamiliar surroundings. A frog will starve to death surrounded by dead flies; for behaviour ceases when imagination cannot replace absent stimuli.

Most machines are controlled by their inputs. A car which does not respond automatically to the steering wheel, the accelerator or the brakes is dangerous. We design most machines to be predictable, so that they respond in expected ways, for then they are generally more useful and safer. But when we consider the rare and unusual machines which do make decisions on their own account, this is no longer true. An automatic pilot has several kinds of information fed into it, and it may select a flight course according to a number of criteria. It is possible to make machines which will play chess, and they may surprise their designer. In short, if the machine is capable of dealing with problems it seems that its inputs must not control its output directly: but by predicting, it will be partly unpredictable.

13 Eye and brain machines

The first edition of this book, which appeared in 1966, just before the current surge of work on Artificial Intelligence, ended with the words: 'We think of the brain as a computer, and we believe that perceiving the world involves a series of computer-like tricks, which we should be able to duplicate, but some of the tricks remain to be discovered and, until they are, we cannot build a machine that will see or fully understand our own eyes and brains.' During the ten years since this was written much has happened in attempting to make machines think and see. Several biologists working on perception of animals and men (including the author) changed their affiliations, to design Robots. This became a new science: Artificial Intelligence (AI) or, as it is sometimes called, Machine Intelligence. This activity sprang from the cybernetics of twenty years before, which produced control and information-handling devices of the greatest importance. They were not however autonomous; they did not make their own basic decisions. Automatic pilots responded to changing conditions to keep a plane in level flight, or land from marker beacon signals; but there were no devices to recognise objects, to appraise situations, or find original solutions to problems. If, indeed, perception is a problem-solving activity, they were far from the philosophical target of a perceiving machine. The difficulties of achieving this proved to be immense. Some of the biological renegade emigrés into 'intelligence technology' (including the author) returned to mice and men. Some administratively influential comments (such as the Lighthill Report, 1973) have attacked Artificial Intelligence as being not only difficult but impossible. Is it impossible? Given that eyes and brains work, why should it be *impossible* to make artificial intelligence? Is there something intrinsically unique about brains?

The optics of eyes was copied in the tenth century. The Arab scholar Alhazen (*c.* 965–1038) realised that Euclid (in his *Optics* of about 300 BC) was wrong in thinking of the eye as a geometrical point (with rays of light shooting out towards objects) appreciating for the first time that it represents objects with images. Alhazen devised the first *camera obscura*. Curiously, he did not report that its images were inverted; but otherwise his description is clear enough. The pin hole *camera obscura* was later described by Giovanni Battista della Porta, in his *Natural Magic* of 1558. Ten years after, the effects of replacing the pin hole with a lens, giving images sufficiently bright for artists to see form and colour clearly, was described by Damielo Barbaro, in his *Practica della Perspectiva* (1568–69). He wrote:

Close all the shutters and doors until no light enters the *camera* except through the lens, and opposite hold a sheet of paper, which you move forward and backward until the scene appears in the sharpest detail. There on the paper you will see the whole as it really is, with its distances, its colours and shadows and motion, the clouds, the water twinkling, the birds flying. By holding the paper steady you can trace the whole perspective with a pen, shade it and delicately colour it from nature.

The secret was out. The world is seen by the brain from images in the *camera obscuras* of our eyes. So, at last, the eyes could be thought of as devices: mechanisms obeying laws of physics and optics. Copies could be made, in various forms, and incorporated into technology for the use of science and art. It was a remarkably important step philosophically, as well as useful in revealing the significance and nature of perspective. The aim now is to achieve the same for the brain: to understand the structure and function of brains so that they may be described functionally; and replicated rather as the *camera obscura* replicated the essentials of biological eyes, though made from quite different materials.

This has deep philosophical interest. If brain function can, like eye function, be transposed to new materials, it would show that there is nothing special about the *substance* of the brain. There is still however some hang over, with a resultant hang up, from Aristotle's notion of *Essence*. Aristotle supposed that the significant properties of things are in their substances. Thus the brain substance, protoplasm, would be regarded as necessary for life and intelligence. The *camera obscura* showed that a glass lens serves as well, or better, than the living tissues and humours of the living eye for forming images. This was one of the many steps towards isolating and reproducing in other forms functional features of living things: and using them to carry

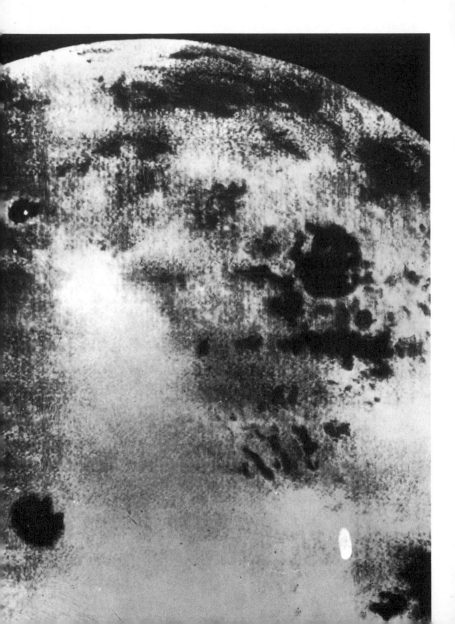

13.1 No eye had ever seen this – the back of the moon, photographed by a Russian rocket in October 1959.

13.2 Nearer than an eye had been — a small region of the front of the
moon, photographed by the American moon probe, Ranger 7, in July 1964.

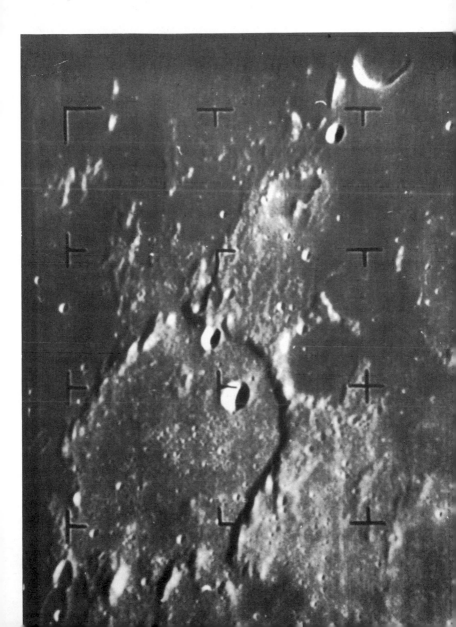

aspects of life into machines. A more dramatic example is the calculating machine, and later the computer.

Devices (including the fingers) have been used for calculation from before the development of the abacus of the sixth century BC. The first mechanism capable of going unguided through steps of arithmetical or logical problems was the calculating engine of Blaise Pascal (1623–62), designed in the year of Newton's birth, 1642.

Gottfried Leibniz (1646–1716) made a more successful calculating engine in 1671, based on the 'Leibniz cylinder', or stepped wheel. It is found in mechanical calculators of the nineteenth century. This line of development led to the first programmable digital computer of 1823 designed (though never fully completed) by the English mathematician Charles Babbage, (1792–1871). This wonderful machine was controlled by punch tape, previously devised for automatic looms, and it printed out its answers directly without human help. If help was required it rang a warning bell, and it introduced safeguards against errors from human stupidity.

The first electrical digital computer was the Harvard Mark I, completed in 1944, using relays; the first electronic digital computer being the ENIAC (Electrical Numerical Integrator and Calculator), completed in 1946. These were practical realisations of Babbage's concepts and designs for a mechanical computer which would follow procedures set up by programs.

A different line of development followed from the Scottish mathematician John Napier (1550–1617) who not only invented logarithms (in 1614), but also devised an apparatus for calculating with logarithms: 'Napier's bones', first described in 1617. This is an early example of an *analogue* computer. It preceded the first *digital* computer of Leibniz by a few years, but was of course much later than the digital abacus which is pre-historic.

'Analogue' and 'Digital' Computing   Engineers often distinguish between *analogue* and *digital*, by whether the device works continuously or in steps. This is obviously an important engineering distinction; but the terms can reflect a deeper distinction. Analogue devices represent functions by copying them. Digital devices represent the symbolic steps by which answers are derived by mathematics or logic. Digital devices follow procedures which we have discovered by analysis; while analogue devices provide answers with no analytical insight required for their design or operation. Arithmetic and logic are not built into analogue devices. They do not have to follow, or carry out the *steps* of arithmetic or logic. Digital computers work in the steps representing each stage of the procedures of mathematics or logic.

Action potentials represent intensities of signals by rate of firing of nerves. So, on the engineers' distinction between 'analogue' and 'digital' they would be classified as digital. But their discontinuous character is not the essential point. They are not *counted* by the synapses, so it is better to think of action potentials as representing intensities by the *analogue* of their rate of firing. On the other hand, when we consider the computing that must go on to make use of sensory signals for deriving perception and producing appropriate command signals for skilled behaviour, then it is very difficult to know whether the brain employs analogue or digital coding. It could of course by a hybrid mixture, Which it is affects how we should interpret electrophysiological data recorded from brain cells. A digital system offers extreme difficulties here, for it would be necessary to record simultaneously from very many cells and know the logical processing rules they are following, before we could 'read' the significance of the recorded activity. Interpreting analogue signals is usually far easier, for they reflect directly the input-output characteristics of the system.

This may seem like a lot of irrelevant detail for considering Eye and Brain; but it can be illuminating to take a more distant view. Regarding physiological mechanisms as applied physics – as bio-engineering – may help in this way. There are, however, those who consider that engineering is too narrow, or may lack appropriate concepts, for describing brain function. This could well be true; but engineering concepts may yet be of some help – and would surely be useful if machines capable of intelligence and perception could indeed be built. It is only engineering which offers this promise. If realised, it will provide solutions to ancient philosophical questions.

Even then we must be careful. Calculating machines are of many kinds; but not all can be like brains. It is for example most unlikely that brains adopt logarithms (or incorporate Napier's bones) though these are remarkably effective for much calculation. It is certain that there are no toothed wheels, or Leibniz cylinders rotating in our heads when we add numbers. There are no transistors either; but nevertheless it is entirely possible that brain cells, or groups of cells, have properties essentially the same as the important properties of transistors, or even of the Leibniz toothed cylinders of early mechanical calculators.

To go back to our potted history of the computer: it is extremely interesting that the design principles of Babbage's wheeled digital computer were transferred without essential change into the electric relays of the Harvard Mark I and into all later electronic computers, using first valves ('tubes') and then solid state components.

What AI claims to be able to do is to repeat more dramatically for biology the move of Babbage's design concepts from wheels to electronics – by transplanting principles of brain function from protoplasm to machines. This would be similar to using the principle of image formation in the eye to produce the *camera obscura*, and later photographic and television cameras. So there are precedents for this kind of move. But is the brain an exceptional case, where functional principles cannot be extracted from its substances, or its particular structures in which they are found in nature? If this were so how could we find out without trying – and for a long time failing? So far we have hardly begun.

Mechanisation of thought

There seems to be a long-standing fear of mechanising thought processes. Helmholtz described perceptions as 'Unconscious Inferences' from sensory data. He arrived at this notion many years after Babbage's calculating machine. (Helmholtz was only two years of age when Babbage's machine was built in 1832.) Although the existence of machine calculation was known for two hundred years (since Pascal) and although it was an issue of popular interest and excitement from Babbage's work of two or three decades before, yet the idea met opposition sufficiently strong for Helmholtz to regret the term 'Unconscious inference'. This 'block' inhibited thinking about the brain and perception in this way, all through the years of Gestalt psychology and the Behaviourism of the first half of the twentieth century. There is resistance still – and this may be limiting our understanding of how we perceive, and think and create. The 'block', if this it is, comes from an ancient belief that consciousness is necessary for logic. This belief persists in the face of wheels carrying out arithmetic, and electronic computers carrying out mathematics so complicated and subtle that few if any mathematicians can fully understand it. Computers are used for designing next-generation computers, and for designing and de-bugging programs. They do not have to be fed with precise instructions; they can be 'self-adaptive', and can learn new skills. They can not only match the world's best checkers player, they can teach him new moves and strategies they have learned by playing games and analysing the games of computers and men. Certainly they must be taught some rules, and what it is to 'win' a game or achieve a 'goal' or a 'solution' – but so also must children be taught these things.

How, then, can it be said that Machine Intelligence is *impossible*? Since *we* are intelligent, for AI to be impossible our brain must be essentially different from any possible machine. As we have seen, the point can hardly be that brains are *conscious* while machines *cannot* be conscious. To establish whether a machine is conscious would undoubtedly be extremely difficult; but it is clear, from what we have already considered, that this is not necessary for we now accept that much of *our* intelligence is not conscious.

The principle objection to Machine Intelligence raised by Sir James Lighthill, in his highly authoritative paper report (page 249) is that a finite machine (a machine having a limited number of states) soon runs out of capacity in problems such as playing chess, because with each move the number of possibilities rapidly increases – to give a 'combinatorial explosion' which would require a machine of astronomical and finally infinite size to encompass. But we – with our brains – can play chess. One is not asking for machines to play *better* chess than we do to demonstrate machine intelligence. If our brains can cope with the combinatorial explosion inherent in the chess situation (and employing data for perception) why should not machines manage also? Surely the spur here is to discover *how brains succeed*; and then ask whether machines could do the same. How do brains succeed? The general answer seems to be that only *likely* possibilities are considered in human thinking and perception. As we have seen (chapter 10) any retinal image is, strictly, infinitely ambiguous. We do not consider or perceive more than a few of these infinite possibilities before arriving at a stable perception. Even the 'Ambiguous Figures' (figures 1.3, 1.4) give only a few alternative perceptions. The 'explosion' is probably avoided throughout the hierarchy from sensory signals to accepted perceptions by, at each crucial step, rejecting unlikely alternatives. This may make our thinking less 'lateral' than some would like; but why should not a machine also be provided with such strategies for its sanity?

The present state of machine perception

Recently robots have been built with television 'eyes' and jointed metal arms and hands, which can perform simple perceptual tasks. They can recognise and pick up objects, and put parts together to make simple models, such as wooden cars from body parts and wheels and axles, which they have correctly selected from a random heap. Developing these machines has proved remarkably difficult. Early estimates were over-optimistic. It has taken a decade of intensive and expensive work by several laboratories to achieve this modest result, and to show that more is possible.

The first kind of object chosen for machine recognition was the cube. The machine learned the various perspective transformations of cubes, then (with its single eye) it could infer the orientation of the cubes, and pick them up and stack them to form towers as a child might do. If a cube-like object was presented, the machine would process its sensed signals as though the object's angles were right angles and its sides of equal length. What it accepts as 'cube' data are then inappropriate for the object being recognised and handled, for its angles are not right angles and its lengths are not all equal – so systematic errors are generated. This is very much our view of visual distortion illusions (chapter 9). The machines' check procedures are less sophisticated than ours, so their behaviour is less reliable and more easily upset: in this sense they suffer more illusions than we do. For example they are easily fooled by shadows crossing the scene, for the edges of shadows are confused with edges of objects. This is rare in human perception; though we did find just this in the case of S.B., when his sight was restored after fifty years of blindness (chapter 11).

Object recognition by television camera and computer is slow even by comparison with young children. It has turned out that the fastest (single channel) digital computers are too slow to go through the steps of their programs to reproduce the speed of biological perception. This may suggest that the brain adopts different strategies: that it works by analogue means, or has rich parallel processing facilities so that many processing problems can be solved simultaneously.

It has turned out that edges of objects are inadequately represented by video signals providing contour information. Also, edges of more distant objects are often partially hidden by nearer objects and so have gaps (see page 182). So the computer, like us, has to use incomplete data to infer the presence of complete objects. This leads to 'line repairing' problems, which make up deficiencies in terms of what ought to be present though not signalled. These programs are hierarchical, starting with line repair and going on up to rejecting unlikely combinations of features and objects. A consequence of such programs is that if a line is *actually* broken, the program may 'repair' it – to invent features which are not present. Surely we see just this with the illusory contours (see figure 5.14). For the computer, as for us, things can be ambiguous (figure 1.3), and they can be paradoxical (figure 12.2). In these shared illusions we see something of our hidden selves reflected in these machines.

This work continues, though perhaps some of the steam has gone out of it while we wait for more suitable 'hardware' for Robot vision. Lighthill, in his report mentioned above, surely correctly pointed out that hardware limitations may be too severe at the present time but this is changing rapidly with the extraordinary progress of computer technology.

The electronic and mechanical hardware of computers and Robot machines are undoubtedly very different from those of natural nervous systems. Computer programs must be matched to the physical possibilities of the computers they are to serve, so we should expect differences between machine and brain programs, if only for this reason. These differences may not however be profound. If machines can be built to display high-level intelligence and efficient perception we have made an immense advance – for then we know at least one solution to how it *can* be done, even if brains are rather different. As it is, present computer technology gives an extremely useful test-bed for checking and suggesting theories of how physiological processes confer perception on eyes and brains.

Machine Perception is not as impressive as we would like. Programs are developing fast, and much has been learned of scene analysis; but it must be admitted that performance does not justify the claim that Machine Perception matches even the modest biological perception given by simple brains. This is embarrassing. The most important point seems to be that organisms are richly endowed with knowledge about the world in which they live. To make effective use of data, there must be appropriate assumptions of the way things are. The problem is to build in knowledge and allow machines to learn about the structure of the world, so that they can develop internal representations adequate for finding solutions within themselves. When machines have knowledge they can be imaginative. When they have imagination they, like us, may live with reality by acting on their postulated alternatives rather than as slaves to signalled events.

Brain states and consciousness

We can find no good reason for pessimism over Machine Perception and Machine Intelligence, though this is a long, difficult and expensive project. There seem to be no objections *in principle* to unconscious inference: it exists in organisms and some machines. Although we may have no need to endow machines with consciousness (or to suppose that they are conscious if they can reason or recognise things) yet it would be nice to know why we are conscious – what purpose consciousness serves. It must be admitted that we do not know. It may however be interesting to end with some philosophical possibilities.

René Descartes held that mind and matter are essentially different. What are the grounds for this distinction – still commonly assumed? Perhaps the main reason is that they *seem*, they *look* very different. Objects are solid, persisting in time, while conscious states are fleeting, insubstantial and do not occupy physical space. Descartes held however that one object – the pineal gland of the brain – is both an object and our internal world. Here the senses come together – to form 'common sense' – in a lump of matter. So if this can happen in the pineal gland, why not in other parts of the brain – or in machines? The pineal gland story is no longer with us; but the central issues remain much as Descartes left them.

There is a recent impetus in philosophy to attack Cartesian dualism in favour of supposing that there is an *identity* between some physical brain states and consciousness; rather as a table is a hard solid object as sensed *and* a bunch of electrons as known to physics. J. J. C. Smart, in particular, has questioned the classical reasons for defending dualism, in an interesting paper: 'Sensations and Brain Processes', appearing in various forms since 1959. Considering an after-image, Smart points out that it is fallacious to say (as some philosophers have said) that because the after-image appears 'out there' in space, though the brain process responsible is in the head, therefore they cannot be the same. For, he points out, the *sensation* is not 'out there'. Sensation could always be in the brain – as aspects of physical brain states – though they *refer* to objects distant in space or time.

Another distinction often made between brain processes and consciousness is that objects (including brains) are *public* while experience is *private*. Will we ever discover enough about brain processes to find physical criteria for recognising consciousness?

A common philosophical objection to a brain-mind identity theory, is that we *speak differently* about physical objects and about conscious experiences. It is often held that because our language behaves differently, there must be basic differences. But does this hold up? Why should there be such wisdom built into the structure of language? The identity theory of consciousness, as held by several present-generation philosophers, especially in Australia, has the interesting characteristic that it flouts the linguistic philosophical tradition: suggesting that this is an empirical question with an empirical answer. The identity relation is not evident in the normal language of how we talk about matter and mind – indeed we describe them as very different – but nevertheless there may be a contingent (not logically necessary) identity. It is an empirical hypothesis that there is an identity between brain states and consciousness. So here we find philosophers joining scientists, asking for experimental evidence rather than philosophical analysis of our language. An identity theory in which consciousness is a property of physical brain

13.3 An object-recognizing and handling robot. Developed in the Department of Machine Intelligence and Perception and subsequently in the School of Artificial Intelligence, Edinburgh University.

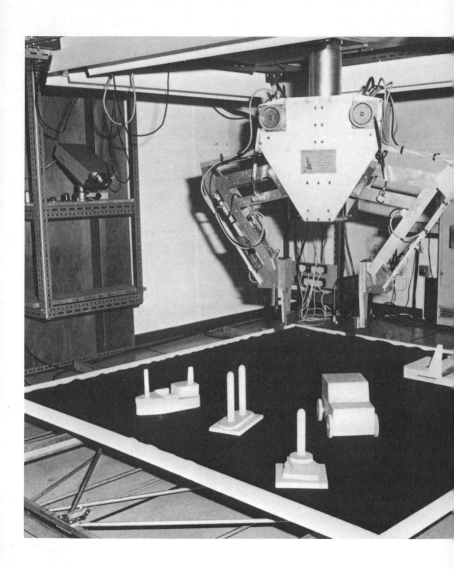

states fits the evidence that minute physical changes of the brain have closely related psychological consequences; but the identity theory is not generally accepted by philosophers or scientists.

Why, we might ask, should it be only the substance or structure of *brains* which have this added property of consciousness? This is different from, for example, tables being both hard and solid *and* also a bunch of electrons in violent motion. This difference-with-identity applies to *all* physical objects, while the supposed brain/mind identity applies only to the substance or structure of *brains*. So we should go on to ask why brains are unique in combining matter with mind. The alternative is to suppose that all matter is to some degree conscious – but we have no evidence at all that stones and tables are aware.

If consciousness turns out to be related to processes giving intelligence, we might however be pushed into thinking of intelligent machines as conscious. It seems highly unlikely that certain *substances* present in brains must be present in machines for them to be intelligent. It is as difficult to conceive 'intelligent substance' as it is to conceive 'conscious substance' – but at least we know that we can make mechanisms carry out processes giving intelligent solutions. This suggests to me that the key to consciousness may be *processes of intelligence* – which might be transferred to machines made of different substances from the brain.

We are left with a very tricky question: does consciousness *affect* physical brain states? (And would consciousness in a machine affect its electronics?) If consciousness does affect physical states, then physiological explanations of perception, intelligence and behaviour cannot be adequate. If on the other hand consciousness does not affect matter, what use can it be, what function can it serve? This seems an insoluble problem if we restrict 'use' and 'function' to physical events: but perhaps this view is too limited for understanding consciousness.

Consciousness is indeed difficult to think about. But for the physicist, matter and force and time, and any other unique attributes of the world, are just as difficult. Science is good at discovering and describing relations; but it is powerless with unique cases. The question 'what is matter?' is as difficult to answer as the question 'what is consciousness?' To suggest that matter is substance is mere tautology. Physics describes matter in terms of structures. It does not ask the ultimate question: 'what are the atomic structures of matter made of?' Similarly, it seems meaningless to ask: 'what is the substance of mind?' When we know enough to describe brain processes capable of giving perception in machines, the problem of consciousness may be solved as the problem of matter is solved in physics: then we need not be ashamed of our ignorance.

Bibliography

If a book has been published both in the United Kingdom and North America both publishers are listed, the British one being listed first. Dates, except where otherwise stated, are of first publication. References to journals are abbreviated in accordance with the *World List of Scientific Periodicals*.

General books

D. C. Beardslee and M. Wertheimer (eds.), *Readings in Perception*, van Nostrand (1958); E. G. Boring, *Sensation and Perception in the History of Experimental Psychology*, Appleton-Century-Crofts (1942); D. E. Broadbent, *Perception and Communication*, Pergamon (1958); J. S. Bruner et al., *Contemporary Approaches to Cognition*, O.U.P./ Harvard (1957); T. N. Cornsweet, *Visual Perception*, Academic Press, 1970; J. J. Gibson, *The Perception of the Visual World*, Allen and Unwin/Houghton Mifflin (1950); J. J. Gibson, *The Senses Considered as Perceptual Systems,* Boston, Houghton Mifflin (1966); R. L. Gregory, *The Intelligent Eye*, Weidenfeld and Nicolson (1970); R. L. Gregory, and E. H. Gombrich, *Illusion in Nature and Art*, Duckworth (1974); R. N. Haber (ed.), *Contemporary Theory and Research in Visual Perception*, Holt, Rinehart and Winston, N.Y. (1968); D. O. Hebb, *The Organization of Behaviour*, Chapman and Hall/Wiley (1949); H. von Helmholtz, *Handbook of Physiological Optics* (1867), ed. J. P. C. S. Southall, Dover reprint (1963); J. E. Hochberg, *Perception*, Prentice Hall (1964); H. W. Leibowitz, *Visual Perception,* Collier/ Macmillan (1965); U. Neisser, *Cognitive Psychology*, Appleton-Century-Crofts (1967); M. H. Pirenne, *Vision and the Eye*, Chapman and Hall-Anglobooks (1948); M. D. Vernon, *A Further Study of Visual Perception*, C.U.P. (1952); L. Zusne, *Visual Perception of Form*, Academic Press, N.Y. (1970).

1 Seeing

For a collection of the classical papers by the Gestalt psychologists: W. H. Ellis, *Source Book of Gestalt Psychology*, Routledge and Kegan Paul (1938).

Roman pavements often have visually ambiguous designs; but spontaneous depth reversal was first described in the last century by a Swiss naturalist who was surprised by sudden changes in appearance of his drawings of crystals: L. A. Necker, 'Observations on some remarkable phenomena seen in Switzerland; and an optical phenomenon which occurs on viewing of a crystal or geometrical solid', *Phil. Mag.* (3 ser.) **1**, 329–37 (1832). This is the origin of the Necker Cube.

2 Light

Sir William Bragg, *Universe of Light*, Bell-Clarke, Irwin (1962); F. A. Jenkins & H. E. White, *Fundamentals of Optics*, McGraw-Hill, 3rd ed. (1957). For a fascinating historical account: Vasco Ronchi, *The Nature of Light*, Heinemann (1970) (trans. of *Storia della Luce* [1939]). Sir Isaac Newton, *Opticks*, 4th ed. (1730), Dover Publications Reprint (1952).

3 In the beginning

There is no book describing primitive eyes in detail, but more developed eyes are described magnificently in G. L. Walls, 'The vertebrate eye and its adaptive radiation', *Cranbrook Institute of Science Bulletin* **19** (1942).

The eyes of insects are described by V. B. Wigglesworth in *The Principles of Insect Physiology*, Methuen/Wiley, 5th ed. (1953). Investigation of the eye of *Copilia* is described by R. L. Gregory, H. E. Ross and N. Moray, in 'The Curious eye of *Copilia*', *Nature* (Lond.) **201**, 1166 (1964). For fairly recent research papers: C. G. Bernhard (ed.), *The Functional Organization of the Compound Eye*, Wenner-Gren Symp. Ser. **7**, Pergamon (1965).

4 The brain

The structure of the brain and nervous system is given in: M. L. Barr, *The Human Nervous System: An Anatomical Viewpoint*, Harper, 2nd ed. (1974); A general discussion of structure and function is given by C. U. M. Smith, *The Brain, Towards an Understanding*, Faber and Faber (1970). Interesting accounts are, D. E. Wooldridge, *The Machinery of the Brain*, McGraw-Hill (1963), and S. Rose, *The Conscious Brain*, Weidenfeld and Nicolson (1973).

Still worth reading today is the classic: Sir Charles Sherrington, *The Integrative Action of the Nervous System* (1906), Yale paperback (1947). The history of ideas on the brain and sensation is given by K. D. Keele in *Anatomies of Pain*, Blackwell/Machwith (1957). The action potential in nerves is described in B. Katz, *Nerve, Muscle and Synapse*, McGraw-Hill (1966).

The important work on discovering the neural mechanism responding to angles of lines and movements, for the cat and monkey brain, is due to D. H. Hubel and T. N. Wiesel, 'Receptive fields, binocular interaction and functional architecture in the cat's visual cortex', *J. Physiol.* **160**, 106 (1962), and many later papers in the same and other journals. For pioneer work on the retina of the frog, see J. Y. Lettvin, H. R. Maturana, W. S. McCulloch, and W. H. Pitts, 'What the Frog's Eye tells the Frog's Brain', *Proc. Inst. Radio Engrs. N.Y.* **47**, (1959). For an interesting discussion of this new physiology: C. Blakemore, 'The Baffled Brain' in *Illusion in Nature and Art* (1974) *op. cit.* For recent experiments suggesting that feature detectors are developed by visual stimulation: C. Blakemore and G. G. Cooper, 'Development of the brain depends on the Visual Environment', *Nature*, **228**, 477–8 (1970) and for a review of this controversial topic see: H. B. Barlow, 'Visual experience and cortical development', *Nature* (Lond.) **258**, 199–204 (1975). 'Conditional after-effects' were first described by Celeste McCollough, 'Colour adaptation of edge-detectors in the human visual system', *Science*, **149**, 1115–16 (1965).

Basic physiology and plasticity of the central nervous system is discussed by J. Z. Young, *The Memory System of the Brain*, O.U.P. (1966); R. Mark, *Memory and Nerve Cell Connections*, O.U.P. (1974); R. M. Gaze, *The Formation of Nerve Conections*, Academic Press (1970); G. A. Horridge, *Interneurons*, W. H. Freeman, N.Y. (1968). For a recent collection of research papers on related topics: M. S. Gazzaniga and C. Blakemore, *Handbook of Psychobiology*, Academic Press, N.Y. (1975).

5 The eye

For the general structure of the eye, see T. C. Ruch and J. F. Fulton, *Medical Physiology and Biophysics*, Saunders, 18th ed. (1960). It is described in great detail in H. Davson (ed.), *The Eye*, Academic Press (1962).

How the eye accommodates to different distances is interesting because the image is the same whether the eye is accommodated too far or too near, so there is no available signal for giving the sign of an error. This has been investigated with an ingenious technique by: F. W. Campbell and J. G. Robson, 'High-speed infra-red optometer', *J. Opt. Soc. Amer.* **49**, 268 (1959).

The full story of the control of the size of the pupil by intensity of light is complicated. See F. W. Campbell and T. C. D. Whiteside, 'Induced pulilary oscillations', *Brit. J. Ophthal.* **34**, 180 (1950). For a fuller account see L. Stark, 'Servo analysis of pupil reflex', *Medical Physics*, ed. O. Glasser, Year Book, Chicago (1960).

The retina is described most generously in *The Retina* by S. L. Polyak, C.U.P./Chicago (1941).

Eye movements were first investigated by R. Dodge, 'An experimental study of visual fixation, *Psychol. Monogr.* **8**, No. 4 (1907). Eye movement control is described in 'Central control of eye movement', by E. Whitteridge, in *Handbook of Physiology – Neurophysiology*, Vol. 2. Optical stabilization of retinal images is first described by R. W. Ditchburn and B. L. Ginsborg, 'Vision with a stabilized retinal image', *Nature* (Lond.) **170**, 36–7 (1952); and by L. A. Riggs, E. Ratliff, J. C. and T. N. Cornsweet, 'The disappearance of steadily fixated visual test objects', *J. Opt. Soc. Amer.* **43**, 459 (1953).

A simple stabilizing technique is described by R. M. Pritchard in 'A collimator stabilising system for the retinal image', *Quart. J. Exp. Psychol*, **13**, 181 (1961). When stabilised, perception fragments in more or less 'meaningful' parts: See R. M. Pritchard, W. Heron and D. O. Hebb, *Canad. J. Psychol.* **14**, 47 (1960). Patterns of eye movements are discussed by the Russian psychologist A. L. Yarbus, *Eye Movements and Vision* (trans. L. A. Riggs); and by R. W. Ditchburn *Eye-Movements and visual perception*, O.U.P. (1973). Basic phenomena of binocular vision are described by K. N. Ogle, *Researches in Binocular Vision*, Saunders (1950). The technique and experiments with cross-correlated random dot patterns to give depth is described by B. Julesz, *Foundations of Cyclopean Perception*, University of Chicago Press (1971).

It is interesting that the random dot pairs do not produce depth when presented with colour contrast but no brightness difference between dots and their background: C. Lu and D. H. Fender, 'The interaction of colour and luminance in stereoscopic vision', *Investigative Ophthal.* **11**, 6 (1972). For a discussion of illusory (or 'cognitive') contours: R. L. Gregory 'Cognitive Contours', *Nature* (Lond.) **238**, 51–2 (1972). For evidence of stereoscopy given by pairs of illusory contours, R. L. Gregory and J. P. Harris, 'Illusory contours and stereo depth', *Perception and Psychophysics* **15**, 3, 511–16 (1974).

6 Seeing brightness

Until fairly recently the accepted theory of dark-light adaptation is that of Selig Hecht: 'The nature of the photo receptor process', in C. Murchisson (ed.), *Handbook of General Experimental Psychology*, O.U.P./Clark U.P. (1934). Doubt was cast on the completeness of this theory by many experiments, including those of K. J. W. Craik: 'The effect of adaptation on differential brightness discrimination', *J. Physiol.* **92**, 406 (1938); and K. J. W. Craik and M. D. Vernon, 'The nature of dark adaptation', *Brit. J. Phychol.* **32**, 62 (1941). It had been greatly modified by Rushton: see W. A. H. Rushton and F. W. Campbell, 'Measurement of rhodopsin in the living human eye', *Nature* (Lond.) **174**, (1954), and many later papers.

Lateral inhibition in the retina is discussed by S. W. Kuffler, 'Discharge patterns and functional organisation of mammalian retina', *J. Neurophysiol.* **16**, 37 (1953); and, for the frog's retina, H. B. Barlow, 'Summation and inhibition in the frog's retina', *J. Physiol.* **119**, 69 (1953). It is discussed in relation to other visual functions, by H. B. Barlow, 'Temporal and spatial summation in human vision at different background intensities', *J. Physiol.* **141**, 337 (1958).

The Pulfrich Effect was first described by E. Pulfrich in *Naturwissenschaften* **10**, 569 (1922) and is discussed by G. B. Arden and R. A. Weale, 'Variations in the latent period of vision', *Proc. Roy. Soc. B.* **142**, 258 (1954).

There is an enormous literature on the absolute sensitivity of the eye. The classical paper on the quantal efficiency of the eye is S. Hecht, S. Schlaer, and M. H. Pirenne, 'Energy quanta and vision', *J. Gen. Physiol.* **25**, 819 (1942). Their method of estimating the number of quanta required for detection using frequency-of-seeing curves is described clearly by M. H. Pirenne, *Vision and the Eye* (chapters 6, 7, and 8), Chapman and Hall/Anglobooks (1948), and T. N. Cornsweet *op. cit.*

The work on recording from the optic nerve of *Limulus* is mainly due to H. K. Hartline, 'The neural mechanisms for vision', *The Harvey Lectures* **37** (1942).

The suggestion that visual detection may be limited by neurological noise, was first made by a television engineer: A. Rose, *Proc. Inst. Radio Engrs. N.Y.* **30**, 293 (1942).

The idea has been developed by several investigators, notably H. B. Barlow, 'Retinal noise and the absolute threshold', *J. Opt. Soc. Amer.* **46**, 634 (1956), and 'Incremental thresholds at low intensities considered as signal noise discriminations', *J. Physiol.* **136**, 469 (1957). Psychophysical decision theory dates from W. P. Tanner and J. A. Swets, 'A Decision-Making Theory of Visual Detection', *Psychol. Rev.* **61**, 401–9 (1954). For the approach described here: R. L. Gregory, *Concepts and Mechanisms of Perception*, Duckworth (1974).

7 Seeing movement

For data on thresholds for detecting movement: J. F. Brown, *Psychol. Bull.* **58**, 89 (1961). More sophisticated measurements are given by H. W. Leibowitz, 'The relation between the rate threshold for perception of movement for various durations and exposures' *J. Exp. Psychol.* **49**, 209 (1955).

The stability of the visual world during eye movements is considered by H. von Helmholtz, *op. cit.* For the 'outflow' theory see E. von Holst, 'Relations between the central nervous system and the peripheral organs', *Brit. J. Anim. Beh.* **2**, 89 (1954); and R. L. Gregory, 'Eye movements and the stability of the visual world', *Nature* (Lond.) **182**, 1214 (1958).

Literature on the autokinetic effect is reviewed by R. L. Gregory and O. L. Zangwill, 'The origin of the autokinetic effect', *Quart. J. Exp. Psychol.* **15**, 4 (1963), where evidence for the muscle-fatigue and 'outflow' theory is given.

The waterfall effect is described by A. Wohlgemuth 'On the after-effect of seen movement', *Brit. J. Psychol. Monogr.* **1** (1911), and by H. C. Holland,

The Spiral After-Effect, Int. Monogr. 2, Pergamon. That it is limited to adaptation of the image/retina system is shown by S. M. Anstis and R. L. Gregory, 'The after-effect of seen motion: the role of retinal stimulation and eye movements', Quart. J. exp. Psychol. (1964). The apparent movement known as the phi phenomenon has been discussed mainly by the Gestalt school: M. Wertheimer, Z. Psychol. **61**, 161 (1912) who named the phenomenon. The time-interval-distance relations of the two lights giving apparent movement of a perceptually single light moving from one to the other (Korte's Law) were given by A. Korte, in K. Koffka (ed.), Beitrage zur Psychologische der Gestalt, Kegan Paul (1919). The effect is in fact extremely variable. A good account is to be found in M. D. Vernon (1961) op. cit.

The cognitive effect of induced movement was first investigated by K. Duncker, 'Induced motion', in W. H. Ellis (ed.), Source Book of Gestalt Psychology, Routledge/Harcourt Brace (1938).

8 Seeing colour

A fascinating collection of classical papers is to be found in R. C. Teevan and R. C. Binney (ed.), Colour Vision, Van Nostrand (1961). This contains Thomas Young's classical paper, 'On the theory of light and colours' as well as those of Helmholtz (1867). E. H. Land, 'Experiments in Colour Vision', Sci. Amer. **5**, 84 (1959), is also included.

For experiments on the effect of adaptation on colour matches, see G. S. Brindley, Physiology of the Retina and the Visual Pathway, Arnold/Waverley Press (1960).

9 Seeing illusions

For a good discussion on dreaming, see I. Oswald, Sleeping and Waking, Physiology and Psychology, Elsevier, Amsterdam (1962), and I. Oswald, 'The experimental study of sleep', Brit. Med. Bull. **20**, 70 (1964). Effects of drugs are discussed by A. Summerfield, 'Drugs and human behaviour', Brit. Med. Bull. **20**, 70 (1964), and by S. D. and L. L. Iversen, Behavioural Pharmacology, O.U.P. (1975).

The work of Wilder Penfield on eliciting memories and sensations by electrical stimulation of the brain is described in W. Penfield and L. Roberts, Speech and Brain Mechanisms, O.U.P. (1959), and W. Penfield, The Mystery of the Mind, Princeton (1975). Effects of repeated patterns giving visual disturbance: D. M. MacKay, 'Interactive processes in visual perception', in Sensory Communication, ed. W. A. Rosenblith, M.I.T. Press and Wiley (1961), which is a useful book.

The first important experimental work on estimating size and shape constancy (following the realisation of the problem by Descartes) was undertaken by Robert Thouless. See R. H. Thouless, 'Phenomenal regression to the real object', *Brit. J. Psychol.* **21**, 339 (1931); and 'Individual differences in phenomenal regression', *Brit. J. Psychol.* **22**, 216 (1932). Thouless used a technique involving comparisons between two discs of cardboard placed at different orientations or distances. A different technique, which can be used for measuring constancy during movement, is described by S. M. Anstis, C. D. Shopland, and R. L. Gregory, 'Measuring visual constancy for stationary or moving objects', *Nature* (Lond.) **191**, 416 (1961). Results of this technique are described in R. L. Gregory and H. E. Ross, 'Visual constancy during movement', *Perceptual and Motor Skills* **18**, 3 and 23 (1964).

Historical discussions of visual distortion and illusions are in R. S. Woodworth, *Experimental Psychology*, Methuen/Holt (1938), and more fully: J. O. Robinson. *The Psychology of Visual Illusion*, Hutchinson (1972). One of the first hints of the kind of theory advocated in this book is R. Tausch, *Psychologische Forschung* **24** (1954). The first statement of the theory given here is R. L. Gregory, 'Distortion of visual space as inappropriate constancy scaling', *Nature* (Lond.) **119**, 678 (1963). Recent papers on 'psychological' explanation of the distortions, in terms of lateral inhibition, are R. H. S. Carpenter and C. Blakemore, 'Interactions between Orientations in Human Vision', *Exp. Brain Res.* **18**, 287–303 (1973) and in terms of Fourier filtering characteristics A. P. Ginsberg, 'Psychological Correlates of a Model of the Human Visual System', *Proc. IEEE NAECON*, Dayton, Ohio, 283–390 (1971). The notion of Size Scaling working both 'upwards' from cues and 'downwards' from the prevailing perceptual hypothesis is presented in R. L. Gregory, *The Intelligent Eye* (1970) *op. cit.*, and R. L. Gregory 'The Confounded Eye' in *Illusion in Nature and Art* (1974) *op. cit.* Evidence of visual scaling applying to touch is given by J. P. Frisby and I. R. L. Davies 'Is the haptic Muller–Lyer a Visual Phenomenon?' *Nature* (Lond.) **231**, 5303 (1970). An investigation of distortions in touch is: R. G. Rudel and H.-L. Teuber, 'Department of visual and haptic Muller–Lyer illusion on repeated trails: a study of cross-modal transfer', *Quart. J. exp. Psych.* **15**, 125 (1963). The lack of illusions in primitive people is discussed by M. H. Segall, T. D. Campbell, and M.J. Herskovitz, *The Influence of Culture on Visual Perception*, Bobbs Merrill, N.Y. (1966); and J. B. Deregowski, 'Illusion and Culture', in *Illusion in Nature and Art* (1973) *op. cit.*

The recent finding that the distortion illusions no longer occur when the scaling for distance by perspective and stereoscopy is correct, is given in: R. L. Gregory and J. P. Harris, 'Illusion-Destruction by appropriate scaling', *Perception* **4**, 203–20 (1975).

10 Art and reality

The Ames' demonstrations are described in W. H. Ittleson, *The Ames Demonstrations in Perception*, O.U.P./Princeton (1952). For Gibson's ideas on 'pick up of information from the ambient array': J. J. Gibson, *The Perception of the Visual World*, Allen & Unwin/Houghton Mifflin (1950) and J. J. Gibson, *The Senses Considered as Perceptual Systems*, Boston (1966). The best discussion relating the problems of the artist to what we know of visual perception is E. H. Gombrich, *Art and Illusion*, Phaidon/Pantheon (1960).

The 'impossible' figures are due to L. S. Penrose and R. Penrose, 'Impossible objects: a special type of illusion', *Brit. J. Psychol.* **49**, 31 (1958). These are in fact drawings. Elaborate examples are by the Dutch artist, the late Maurits Escher; see J. L. Locher (ed.), *The World of M. C. Escher*, Harry N. Abrams, N.Y. (1971). See also: Marianne L. Teuber, 'Sources of Ambiguity in the Prints of Maurits C. Escher', *Sci. Amer.* **231**, 1 (1974). Truly three dimensional objects which look impossible from critical points of view are discussed by the author in *The Intelligent Eye* (1970).

For the history of perspective see: J. White, *The Birth and Rebirth of Pictorial Space*, Faber and Faber (1967).

11 Do we have to learn how to see?

For experiments on animals reared in darkness, see E. H. Hess, 'Space perception in the chick', *Sci. Amer.* **195**, 71 (1956); A. H. Reisen, 'The development of perception in man and chimpanzee', *Science* **106** (1947); A. H. Reisen, 'Arrested vision', *Sci. Amer.* **183**, 16 (1950).

For a discussion of cases of human recovery from blindness, up to 1932: M. von Senden, *Space and Sight* (tr. P. Heath), Methuen/Free Press (1960). These cases came into prominence with D. O. Hebb's *The Organisation of Behaviour*, Chapman & Hall/Wiley (1949). A recent case is described by R. L. Gregory and J. G. Wallace, '*Recovery from early blindness: a case study*', *Exp. Psychol. Soc. Monogr. No. 2*, Cambridge (1963) (reprinted in R. L. Gregory, *Concepts and Mechanisms of Perception*, Duckworth [1974]. This contains the full account of the case of S.B. described in this chapter. Recovery of sight with an artificial lens – in a hollow tooth – in the eye is discussed by Alberto Valvo, *Sight Restoration Rehabilitation*, Amer. Foundation for the Blind, 15 West 16th Street, New York, N.Y. 10011 (1971).

For the work of recording eye movements of young babies, see R. L. Fantz, 'The Origin of Form Perception', *Sci Amer.* **204**, 66 (1961) (offprint No. 459).

For G. M. Stratton's work, see 'Some preliminary experiments on vision',

Psychol. Rev. **3**, 611 (1896); 'Vision without inversion of the retinal image', *Psychol. Rev.* **4**, 341 (1897), and *Psychol. Rev.* **4**, 341 (1897). For Ewert's experiments, P. H. Ewert: 'A study of the effect of inverted retinal stimulation upon spatially co-ordinated behaviour', *Genet. Psychol. Monogr.* **7**, 177 (1930); and two papers on 'Factors in space localization during inverted vision', *Psychol. Rev.* **43**, 522 (1936), and **44**, 105 (1937). This was followed up in J. & J. K. Peterson 'Does practice with inverting lenses make vision normal?', *Psychol. Monogr.* **50**, 12 (1938). Further references, and original work especially on displacement of images in time: K. U. and W. M. Smith, *Perception and Motion: an Analysis of Space-structured Behaviour*, Saunders (1962). The important work of Richard Held and his colleagues on adaptation in humans to displacing prisms is given in many papers: see R. Held and A. Hein 'Movement-produced stimulation in the development of visually guided behaviour', *J. Comp. and Phys. Psychol.* **56** (1963) for the experiment described in the text.

The first paper on the effect of distorting spectacles is J. J. Gibson, 'Adaptation, after-effect and contrast in the perception of curved lines', *J. exp. Psychol.* **16**, (1933). Figural after-effects are described in W. Kohler and H. Wallach, 'Figural after-effects', *Proc. Amer. Phil. Soc.* **88**, 269 (1944), and C. E. Osgood and A. W. Heyer, 'A new interpretation of figural after-effects', *Psychol. Rev.* **59**, 98 (1951). For a full review on orientation and adaptations: I. Howard and W. Templeton, *Human Spatial Orientation*, Wiley (1966). See also I. Rock, *Orientation and Form*, Academic Press, N.Y. (1973).

The great pioneer in child development is of course Jean Piaget. There is an enormous literature on his work; perhaps his own most relevant books are: J. Piaget, *The construction of reality in the child*, Routledge & Kegan Paul (1956) (French edition 1937), and *The Child's Conception of Space*, Routledge & Kegan Paul (1956) (French edition 1948).

The most interesting and reliable experimental work on development of perception in infants is by Jerome Bruner. Many of his papers on these and other topics are in: J. S. Bruner, *Beyond the information given*, Allen and Unwin (1974). See also J. S. Bruner and Barbara Kowslowski, 'Visually preadapted Constituents of Manipulatory Action', *Perception* **1**, 3–14 (1972).

Some evidence of early baby recognition is given by Tom Bower: T. G. R. Bower, 'The visual world of infants', *Sci. Amer.* **215**, 80–92 (1966) (offprint No. 502); 'The object in the world of the infant', *Sci. Amer.* **225**, 4 (1971) and 'Object Perception in Infants', *Perception* **1**, 15–30 (1972). The 'visual cliff' experiment is described in: E. J. Gibson and R. D. Walk, 'The Visual Cliff', *Sci. Amer.* **202**, 64–71 (1960) (offprint No. 402). Development particularly of reading is discussed by Eleanor Gibson in: E. J. Gibson,

Principles of Perceptual Learning and Development, Appleton-Century-Crofts (1969).

12 Seeing and believing

For Albert Michotte's work on seeing causality: T. R. and E. Miles, *Perception of Causality*, Methuen (1963).

For philosophical discussions of 'other minds' and related problems: Ludwig von Wittgenstein, *Philosophical Investigations*, Blackwell (1953). J. Wisdom, 'Other Minds', *Proc. Aristotelian Soc. Suppl. 20,* 122–147 (1946). D. M. Armstrong, *A Materialist Theory of Mind*, Routledge & Kegan Paul (1968).

The important notion that scientific observations depend on general assumptions – 'paradigms' such as Darwinian evolution (which is how we see perception) – is discussed admirably by T. S. Kuhn, *The Structure of Scientific Revolutions*, 2nd ed. *Int. Encl. of Unified Science* **2**, Chicago (1970).

For recent philosophical discussion of the Mind–Brain Identity Theory: C. V. Borst (ed.), *The Mind–Brain Identity Theory*, Macmillan (1970).

13 Eye and Brain machines

For a collection of classical papers on Artificial Intelligence: E. Feigenbaum and J. Feldman (eds.), *Computers and Thought*, McGraw-Hill (1963). For a brilliant example of A.I.: T. Winograd, *Understanding Natural Language*, Edinburgh (1972). For a mathematical treatment of pattern recognition: J. R. Ullmann, *Pattern Recognition Techniques*, Butterworths (1973).

For a critical discussion of the philosophy and possibilities of A.I.: Sir James Lighthill, *Artificial Intelligence*, Science Research Council (Lond.), State House, High Holborn, London W.C.1, paper symposium (1973). For a recent 'positive' review, relating A.I. concepts to human perception: A. K. Mackworth, 'Model-Driven Interpretation in Intelligent Vision Systems', *Perception* **5**, 3, 349–70 (1976).

For a discussion of purpose in machines and man: Margaret A. Boden, *Purposive Explanation in Psychology*, Harvard University Press (1972), and for the most recent review: Margaret A. Boden, *Artificial Intelligence and Natural Man*, Harvester Press, Hassocks, Sussex: Basic Books, NY (1977).

Acknowledgments

My interest in perception started with the teaching of Professor Sir Frederic Bartlett, FRS, and was encouraged by Professor O. L. Zangwill, at Cambridge University.

I would like to thank particularly Dr Stuart Anstis, and my other colleagues and students who have discussed the problems of this book with me, pointed out errors, and helped with experiments. I have benefited by the personal generosity of many people in the United States, especially: the late H.-L. Teuber, the late Warren McCulloch and F. Nowell Jones. The book was largely written during a visit to Professor Jones's department at UCLA.

I would like to thank Mrs Audrey Besterman and Miss Mary Waldron for drawing the majority of the diagrams and the Illustration Research Service, London, for collecting the colour plates used.

Acknowledgment is due to the following for illustrations: 1.2 C. E. Osgood and Oxford University Press; 2.1, 10.1 The British Museum; 2.2, 8.1 The Royal Society; 2.4 The Bodleian Library; 3.1 G. L. Walls and *Cranbrook Institute of Science Bulletin*; 3.2 M. Rudwick; 3.3 V. B. Wigglesworth, Methuen & Co., Ltd and John Wiley and Sons, Inc.; 3.3, 3.5 R. L. Gregory, H. E. Ross, N. Moray and *Nature*; 4.2 W. Penfield, T. Rasmussen and The Macmillan Co., New York; 4.4 The British Broadcasting Corporation; 4.6, 4.7 D. H. Hubel, T. H. Wiesel and the *Journal of Physiology*; 5.2 T. C. Ruch, J. F. Fulton and W. B. Saunders Co.; 5.5 R. M. Pritchard and *Quarterly Journal of Experimental Psychology*: 5.16 Bela Julesz and *Science* (Vol. 145, 24 July 1964, pp. 356–62), © 1964 by the American Association for the Advancement of Science; 6.6, 6.7, 6.8 H. K. Hartline and Academic Press, Inc.; 7.2, 9.1 The Mansell Collection; 7.4 R. L. Gregory, O. L. Zangwill and *Quarterly Journal of Experimental Psychology*; 8.3 W. D. Wright and Henry Kimpton; 8.4 S. Hecht, C. Murchisson and Clark University Press; 9.9 Rania Massourides; 9.12, 10.16 Derrick Witty; 9.17 J. Allen Cash; 10.3 Drawings Collection, Royal Institute of British Architects; 10.2 Bibliothèque de l'Institut de France; 10.4 John Freeman; 10.5 by courtesy of the Trustees of the National Gallery, London; 10.8, 10.9, 10.10 *Punch*; 10.11 Eastern Daily Press, Norwich; 10.13 J. J. Gibson, Allen & Unwin Ltd., and Houghton Mifflin Company; 11.1 William Vandivert and *Scientific American*; 11.2, 11.3 R. L. Gregory, J. G. Wallace and *Experimental*

Psychology Society; 11.4 R. L. Fanz and *Scientific American*, photo David Linton; 11.7 I. Kohler and *Scientific American*; 11.11, 11.12, 11.13 K. U. and W. M. Smith and W. B. Smith and W. B. Saunders Co.; 12.1 A. Michotte, Methuen & Co. Ltd., and Basic Books, Inc. Publishers; 12.2 L. S. and R. Penrose and *British Journal of Psychology*; 13.1, 13.2 London and Wide World Photos; 13.3 Ed Morrett and Whifin Machine Co.

R.L.G.

Index

Pitts, W. H., 94
Plato, 15
Poggendorf figure, 139, 140, 144
Polyak, S., 63
Ponzo illusion, 139, 144
Porta, G. B. della, 227
Pritchard, R., 59
proprioceptive adaptation, 212
protanopia, 128–9
pseudoscopic vision, 67, 68
Pulfrich Pendulum effect, 82, 83
pupils, examination of, 56; form and function, 51, 53–6; size, 52–3
Purkinje images, 52
Purkinje shift, 81, 85

quanta, 18, 21–2

rabbits, registering movement, 97
railway lines illusion, 139, 151
random background, 86–91, 92
range finders, 66
ray figures, 135–7
Rayleigh, Lord, 129
reaction-time, 41, 83
reading, perception and, 204
Realist theory of perception, 181
receptors, *see* photoreceptors
red-green colour blindness, 127–9
redundant figures, 135, 136
refraction of light, 17–18, 51
Reissen, R. L., 191
retina, and the brain, 44, 61; brightness and, 78; eye movement and, 58; form and function, 60–4; fovea, 61;

lateral inhibition, 80; peripheral, 61–3, 93; photopigments, 64, 79–80; sensitivity, 21–2; stimulation, 10; *see also* photoreceptors
retinal delay, 83
retinal images, displacement of, 204–5
retinal rivalry, 67–9
rhodopsin, 79
Riley, B., 137
Robots, 42, 226, 234–6
rods, 60, 61–3, 78, 79, 81
Roemer, O., 15
rotating window, Ames, 180–1
Rushton, W. A. H., 78–80

saccades, 56, 58–9
Scheiner, C., 60
schizophrenia, 134
Schlaer, S., 22
scotopic vision, 61
shadows, 166, 182–7, 234
shape constancy, 153, 154
Sheridan, R. B., 106
Sherrington, C., 98, 99
Sine Law, 17
size constancy, 152–7
Smart, J. J. C., 237
Smith, K. U. and W. M., 213, 214
Snell, W., 17
spectacles, 25, 54
spectral luminosity curve, 81
speed of light, 15–17
Sperry, R. W., 209
'spots before the eyes', 49
stereo projection, 116
stereo vision, 64–75
stereoscope, 67
Stratton, G. M., 204–6, 209
stretch receptors, 99–100, 101
stroboscopes, 112
superior colliculus, 46

tears, 29

telestereoscope, 68
television, movement, 111–13
temporal displacement of images, 215, 216, 217
texture gradients, perspective, 181–2
thought, perception and, 14
Thouless, R., 153–4
touch, displaced images and, 209; illusions, 162; and recovered sight, 194–7
Treviranus, 60–1
trilobites, eyes, 29
tritanopia, 128

Valvo, A., 200
velocity, determination, 97
visual cliff experiment, 190, 201–2
visual cortex, 44–9
visual projection area, 45, 97

Wald, G., 64
Wallace, J., 194
waterfall effect, 106–10, 143
wavelengths, light, 18–21
Weber's Law, 88–90
Wheatstone, C., 67
white light, composition, 18
Wiesel, T. N., 46, 47, 108, 143
Wittgenstein, L., 218
Woodworth, R. H., 147
Wright, W. D., 123

yellow, perception of, 121; receptors for, 129–31
Young, T., 118, 120–4, 125

Zöllner figure, 139, 140, 144
zonula, 51
Zulus, circular culture, 161–2